THE NUTS and BOLTS of PROOFS

ANTONELLA CUPILLARI

Penn State—The Behrend College

Wadsworth Publishing Company
Belmont, California
A Division of Wadsworth, Inc.

Mathematics Editor: Barbara Holland
Production Editor: Leland Moss
Designer: James Chadwick
Copy Editor: Margaret Moore
Compositor: Moss/Chadwick/Mac Associates
Print Buyer: Barbara Britton
Technical Illustrator: Alan Noyes
Editorial Assistant: Sally Uchizono

Printed in the United States of America **49**

1 2 3 4 5 6 7 8 9 10—93 92 91 90 89

Library of Congress Cataloging-in-Publication Data

Cupillari, Antonella.
The nuts and bolts of proofs.

Bibliography: p.
1. Proof theory. I. Title.
QA9.54.C86 1989 511.3 88-33905
ISBN 0-534-10320-0

Preface

This booklet is addressed to those who are interested in learning more about how and why proofs of mathematical statements work.

The only background required is the material covered in a first semester calculus class. Therefore most of the statements considered will deal with properties of numbers and basic properties of functions. The purpose of these notes is to help the reader gain a better understanding of the basic logic of mathematical proofs, which does not greatly depend on the level of the course. Indeed the fact that a statement might seem "easy" to understand does not imply that proving it will be an effortless task.

George Polya writes "A great discovery solves a great problem but there is a grain of discovery in the solution of any problem. Your problem may be modest; but . . . if you solve it by your own means, you may experience the tension and enjoy the triumph of discovery."

I would like to thank Mr. Carl R. Knittle and Mr. David Flynn, who read the first version of the manuscript and made many valuable suggestions, and Dr. Robert P. Hostetler for his encouragement.

To my family

Contents

List of Symbols

Natural or counting numbers: $\mathbb{N} = \{1, 2, 3, 4, 5, \ldots\}$

Prime numbers = { a counting number is called prime if it is divisible only by itself and 1} = {3, 5, 7, 11, 13, ...}

Whole numbers = W = {0, 1, 2, 3, 4, 5, ...}

Rational numbers = $\mathbb{Z}$ = {all numbers of the form a/b with a and b integers and $b \neq 0$}

Irrational numbers = {numbers having an infinite decimal part not showing a repeating pattern}

Real numbers = $\mathbb{R}$ = {all rational and irrational numbers}

Complex numbers = $\mathbb{C}$ = {all numbers of the form $a + ib$ with a and b real numbers and i such that $i^2 = -1$}

$n! = n(n-1)(n-2) \ldots (3)(2)1$

$n!$ (read "n factorial") is defined for $n \geq 0$. By definition $0! = 1$.

$a \in A$ a is an element of A (see Equality of Sets)

$a \notin A$ a is not an element of A (see Equality of Sets)

$B \supseteq A$ A is contained (or equal to) in B (see Equality of Sets)

$\{x \mid x$ has a certain property$\}$ description of a set. The symbol | is read "such that."

$A \cup B$ read "A union B" (see Equality of Sets)

$A \cap B$ read "A intersection B" (see Equality of Sets)

$A' = C(A)$ read "complement of A" (see Equality of Sets)

$|x|$ = absolute value of x = distance from 0 to x

$lcm(a,b)$ = least common multiple of a and b

GCD(a,b) = greatest common divisor of a and b

Properties of Real Numbers

Addition

- The sum $a + b$ of two real numbers is a real number.
- Associative property: $(a + b) + c = a + (b + c)$ for all real numbers a, b, c
- Commutative property: $a + b = b + a$ for all real numbers a, b
- Additive identity: There exists a number, 0, such that $a + 0 = a$ for all real numbers a.
- Opposite: For every number a there exists a number, indicated as $-a$, such that $a + (-a) = 0$.

Multiplication

- The product $a\,b$ of two real numbers is a real number.
- Associative property: $(a\,b)c = a\,(b\,c)$ for all real numbers a, b, c
- Commutative property: $a\,b = b\,a$ for all real numbers a, b
- Multiplicative identity: There exists a number, 1, such that $a\,1 = a$ for all real numbers a.
- Reciprocal: For every number $a \neq 0$ there exists a number, indicated as a^{-1}, such that $a\,(a^{-1}) = 1$.

Distributive property $a\,(b + c) = a\,b + ac$ for all real numbers a, b, c

Trichotomy property If a and b are two real numbers, then either $a < b$, or $a = b$, or $a > b$.

Well-Ordering Principle If A is a collection of positive integers, containing at least one number (A is nonempty), then A contains a smallest number.

Introduction and Basic Terminology

Have you ever felt that the words *mathematics* and *frustration* have a lot in common? There are many people who do. This feeling is often the result of the student using an unproductive and often unsystematic (and panicky) approach that leads to hours of tedious work, which leaves the student feeling uncomfortable with the subject. When anxiety sets in, memorization may look like the way to "survival." But memorization without thorough understanding is usually a poor and risky approach, both in the short and in the long run. It is difficult to successfully recall a large amount of memorized material under the pressure of an exam. And once the course is over, most of the material will quickly fall into oblivion. This will render most of the work done completely useless, and it will make future use of the material very difficult.

Proofs represent one of the major obstacles that students face when studying advanced math. Some students prefer to believe the results proved are true without checking them or understanding why. But there is much to be learned from an understanding of the proofs themselves. Such an understanding gives us new techniques that we can use to gain an inside view of the subject, obtain other results, and remember the results more easily or rederive them if we forget them.

To learn how to read and understand proofs already written in a textbook, we will proceed by breaking them down into a series of simple steps and looking at the clues that lead from one step to the next. "Logic" is the key that will help us in this process. We will use the words *logic* and *logical* according to a definition suggested by Irving Copi: "Logic is the study of methods and principles used to distinguish good (correct) from bad (incorrect) reasoning" [1]. Once we are able to recognize the mechanism of proofs already developed, we can try to construct proofs on our own.

Before we start, though, we need to know the precise meaning of some of the most common words that appear in mathematics books.

Statement: A *statement* is a sentence expressed in words that is either true or false.

Statements do not include exclamations, questions, or orders. A statement cannot be true and false at the same time, though it can be true or false when considered in different contexts. For example, the statement "No man has ever been on the moon" was true in 1940, but it is now false.

A statement is *simple* when it cannot be broken down into other statements ("It will rain." "Two plus two is four." "I like that book."). It is *composite* (or compound) when it contains several simple statements connected by words such as *and, although, or, thus, then, therefore, because, for, moreover, however,* and so on ("It will rain and it is windy." "I like that book because it is controversial.").

Definition: A *definition* is a statement of the precise meaning of a word or phrase, a mathematical symbol or concept.

Definitions are like the soil in which a theory grows. It is difficult to understand and work with results that use technical terms if their definitions are not clear to us. Knowing and understanding the definitions will save a lot of time and frustration.

This is not to suggest that definitions should be memorized by rote, without understanding them. One needs to work with new definitions and be sure their meanings and immediate implications are clear, so that it is possible to recall and use them quickly and appropriately. It is easy to fall behind during a lecture when the speaker is using unfamiliar words and to miss much of the speaker's argument because one is either trying to remember the meaning of the words or has already lost interest, not understanding what is being said. In this situation, conscious or unconscious doubts about one's quantitative/mathematical abilities creep in, making successful and efficient learning more difficult.

So we should make sure we have a good starting point by having a clear and thorough understanding of all the necessary definitions. It is usually useful to pin down a definition through some examples that satisfy it and some examples that do not satisfy it (Do not confuse the two concepts, though. Examples are not definitions).

Theorem: A *theorem* is a mathematical statement that can be proved by a chain of logical reasoning on the basis of certain assumptions that are given or implied in the theorem.

The word *theorem* shares its Greek root with the word *theatre*. They both come from *thea,* which means the act of seeing. Indeed, the proof of a theorem usually allows us to see further into the subject we are working on.

Lemma: A *lemma* is an auxiliary theorem proved beforehand to be used in the proof of another theorem.

Some theorems can have proofs that are long and difficult to follow. In such cases it is common for one or more of the main intermediate steps to be isolated as lemmas and proved before proving the theorem. Then, in the proof of the theorem we can refer to the lemmas already established and use them to go to the next step. Often the results stated in lemmas are not very interesting in themselves, but they play key roles in the proofs of more important results.

Corollary: A *corollary* is a theorem that is a natural consequence, that is, it follows logically from something else.

Corollaries can be as important as theorems. The word underlines the fact that the results stated in corollaries follow relatively easily from a main theorem or a proposition already proved. The James & James *Mathematics Dictionary* defines a corollary as "a by-product of another theorem."

General Suggestions

The first step, whether we are trying to prove a result or to understand someone else's proof, consists of clearly understanding what are the assumptions (hypotheses) made in the statement of the theorem and what is the conclusion to be proved. In this way we are establishing the starting and ending points of the thinking process that will take us from the hypothesis to the conclusion.

We must understand the hypotheses so that we can use the full strength of the information given, either implicitly or explicitly, to achieve the desired result.

It is essential to check all the technical words appearing in the statement and to review the definitions whose meanings are not completely familiar *and* clear.

Examples

1. Suppose we are going to prove the following result:

If a triangle is equilateral, then its internal angles are equal.

We start with the following information:

(i) The object is a triangle (explicit information).
(ii) The three sides have the same length (explicit information).

But what else do we know about triangles, that is, what implicit information do we have? Well, we can use any previously proven result, not only about triangles, but also, for example, about geometric properties of lines and angles in general.

2. Consider the following statement:

   Let a be a nonzero real number.

   The statement gives the following information:

   (i) a is different from zero (explicit information).
   (ii) a is a real number (explicit information).

   As mentioned above, the second fact implicitly says that we can use all the properties about real numbers that our textbook has already mentioned or it requires the reader to know.

Sometimes the hypothesis, as stated, might contain some nonessential details, which are given for the sake of clarity.

Examples

1. Consider the triangle ABC.
2. Let A be the set of all even numbers.

   The fact that the triangle is called ABC is not significant. We can use any three letters to name the three vertices. In the same way, we can use any letter to denote the set of all even numbers. The important thing is consistency. If we have used the letters A, B, and C to denote the vertices of the triangle, then these letters will refer to the vertices of the triangle any time they are mentioned in that proof, and they cannot be used to represent other objects.

Only after we are sure that we can identify the hypothesis and the conclusion and that we understand the meaning of a theorem to be proved, can we go on to read, understand, or construct its proof, that is, the logical argument that will establish why the theorem we are considering is true.

At this point we would like to emphasize the difference between the validity of an argument and the truth or falsity of the results of an argument.

An argument is ***valid*** if its hypothesis supplies certain, sufficient basis for its conclusion. An argument can be valid and reach a false conclusion, as in the following example, in which one of the hypotheses is false.

> Penguins are birds.
> All birds are able to fly.
> Therefore penguins are able to fly.

In other cases an invalid argument can reach a true conclusion. Consider the following argument:

> If Joe wins the state lottery, he could afford a new car.
> Joe did not win the state lottery.
> Therefore Joe cannot afford a new car.

It is obvious that this is an example of bad reasoning. Indeed, Joe did not win the state lottery, so he might not be able to afford a new car (the conclusion is true). But, on the other hand, Joe might win the church raffle (or he might already be wealthy) and he will be able to afford a new car (the conclusion is false, whereas the hypotheses are still true). So the conclusion does not logically follow from the hypotheses.

When we are working on a proof, our main concern must be the use of valid arguments, not the elegance of the construction. Very often the construction of a sound proof takes considerable time and effort. Thus we must be ready to write more than one draft.

The suggestions given in this section can be used as a guide when tackling any theorem.

Some Basic Techniques Used in Proving a Theorem of the Form "A Implies B"

Let's start by looking at the details of a process that goes on almost automatically in our brains hundreds of times every day: deciding whether a "statement" is true or false.

Suppose that you make the following statement: "If I go home this weekend, then I will take my parents out to dinner."

When is your statement true? When is it false?

The statement we are considering is a composite statement, and its two parts are the simple statements:

A: I go home this weekend.
B: I will take my parents out to dinner.

As far as your trip goes, there are two possibilities:

1. You are going home this weekend (A is true).
2. You are not going home this weekend (A is false).

Regarding the dinner, we can examine two possibilities as well:

1. You will take your parents out to dinner (B is true).
2. You will not take your parents out to dinner (B is false).

So we can consider four possibilities:

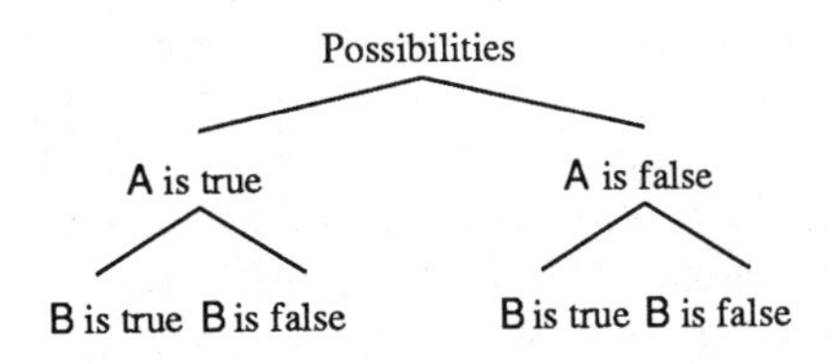

If you did not go home, nobody can accuse you of lying because you did not take your parents out to dinner. Nor can you be accused of lying if you do not go home for the weekend, but you took your parents out to dinner, since they came to visit you. (In these two cases A is false.)

If you do go home and take your parents out to dinner, your statement is true.

Thus there is only one case in which you could be accused of having lied: You went home for the weekend, but you did not take your parents out to dinner.

In conclusion, there is only one case in which your statement is false, namely, when A is true, while B is false.

This is a general feature of statements of the form "if A, then B " or "A implies B."

> *A statement of the form "if A, then B" is true if we can prove that it is impossible that A is true and B is false at the same time.*

In other words, this principle says: If we assume that "A implies B" and that A is true, we must conclude that B is true.

(All arguments having this form, which is called ***modus ponens***, are valid. The expression "modus ponens" comes from the Latin *ponere*, meaning "to affirm.")

The statement "if A, then B" is equivalent to the statement "A is a sufficient condition for B" and to the statement "B is a necessary condition for A." (Consult the James & James *Mathematics Dictionary* if you would like to find out more about "sufficient" and "necessary" condition.)

Since in a statement of the form "if A, then B" the hypothesis and the conclusion are clearly separated (part A, the hypothesis, contains all the information we are allowed to use; part B is the conclusion), we prefer to write theorems in this form.

The following steps can make the statement of the theorem simpler and therefore more manageable, *without changing its meaning*.

1. Identify the hypothesis (A) and conclusion (B), so that the theorem can be written in the form "A implies B" or "if A, then B."
2. Watch out for irrelevant details.
3. Rewrite the statement to be proved in a form you are comfortable with.
4. Check all the relevant properties (from what you have studied up to this point) of the objects involved. If you get stuck during the proof, double-check whether you have overlooked some explicit or implicit information you are supposed to know and use.

The examples in the next section will show us how to use the preceding suggestions to prove whether some statements are true.

DIRECT PROOF

A direct proof is based on the assumption that the hypothesis is true and proceeds by a series of logically connected steps to the conclusion.

Example 1: The sum of two prime numbers larger than 2 is not a prime number.

Discussion: The statement is not in the standard form "if A, then B." Therefore we have to identify the hypothesis and the conclusion.

What explicit information do we have? We are dealing with any two prime numbers larger than 2.

What do we want to conclude? We want to conclude that their sum is not a prime number. So we can set:

A: Take any two prime numbers larger than 2 and add them.
(*Implicit hypothesis:* Since prime numbers are natural numbers, we can use the properties and operations of natural numbers.)

B: Their sum is not a prime number.

Thus we can rewrite the statement as: If we take any two prime numbers larger than 2 and add them, we will obtain a number that is not prime. This statement is less elegant than the given one, but it is more explicit.

What is a prime number?

A counting number is called prime if it is divisible only by itself and 1 (1 is not a prime number). For example, the numbers 2, 3, 5, 7, 11, 13, 17, 19, 23 are prime.

What do all the prime numbers larger than 2 have in common? Why was 2 discarded?

If we look at the numbers we have written, we could be tempted to conclude that all the prime numbers larger than 2 are odd. Let's be extremely careful about this point. We can see that a few prime numbers are odd. But what about all the other prime numbers? We cannot generalize from a few examples because we might risk *overgeneralizing* (generalizing that results in obtaining a false statement).

Could we conclude, using stronger evidence, that all the prime numbers larger than 2 are odd? Since a prime number larger than 2 is divisible only by itself and 1, it is not divisible by 2. So it is not even. Thus a prime number larger than 2 must be an odd number.

What happens if we add two odd numbers larger than 2? The sum of two numbers larger than 2 is a number larger than 2. From experience we know that the sum of two odd numbers is an even number. But this is not good enough evidence (overgeneralizing, again). We must prove this fact.

Let a and b be two odd numbers larger than 2. Since they are odd, they are not divisible by 2. Therefore, when we divide them by 2, we get a remainder. The remainder of a division is a nonnegative number smaller than the divisor. Thus the remainders of these divisions by 2 must be 1. So we can write

$$a = 2t + 1 \quad \text{and} \quad b = 2s + 1$$

with t and s positive integers. So

$$a + b = (2t + 1) + (2s + 1) = 2t + 2s + 2 = 2(t + s + 1)$$

with $t + s + 1 > 1$.

This sum is therefore divisible by 2. So it is not a prime number. We reached the conclusion contained in the original statement! We are on the right track. Can we write the proof in a precise and "clean" way? Let's try!

Proof: Let n and p be two prime numbers with $n > 2$ and $p > 2$.

Since the numbers are prime and larger than 2, they are not divisible by 2. Therefore, when we divide them by 2, we get a nonzero remainder. The remainder of a division is a nonnegative number smaller than the divisor. Thus the remainders of these divisions by 2 must be 1. So we can write

$$n = 2q_1 + 1 \quad \text{and} \quad p = 2q_2 + 1$$

where q_1 and q_2 represent the quotients of the divisions. Thus

$$n + p = (2q_1 + 1) + (2q_2 + 1) = 2(q_1 + q_2 + 1) = 2t$$

with $q_1 > 0$ and $q_2 > 0$.

Since $t = q_1 + q_2 + 1$ is an integer larger than 1, $n + p$ is a multiple of 2. (This step shows that the sum of two odd numbers is an even number.) This implies that $n + p$, the sum of two prime numbers, is not a prime number.

■

How does this proof relate to the considerations at the beginning of this section?

We have worked under the assumption that part A of the statement is true. We have shown that part B holds true (and we have not done this through an example, but through a general way of reasoning; we will talk about this distinction in detail later on). Therefore it is true that A implies B.

Let's consider another proof.

Example 2: Let f and g be two real-valued functions defined for all real numbers. If f and g are one-to-one, then $f \circ g$ is a one-to-one function.

Discussion:

A: We are considering two functions, f and g, which have the following properties:

1. They are defined for all real numbers.
2. They are one-to-one.

(The fact that the functions are indicated by f and g is irrelevant. We can use any two letters.)

B: The function $f \circ g$ is a one-to-one function.

We can understand the meaning of the given statement if we are familiar with the definitions of function defined on real numbers, of one-to-one function, and of composition of functions.

By definition, a real-valued function h (do not use f and g, since these letters are already in use) is said to be one-to-one if for any two distinct real numbers x_1 and x_2 (that is, $x_1 \neq x_2$)

$$h(x_1) \neq h(x_2).$$

By definition of composition of functions, the function $f \circ g$ is defined as

$$f \circ g(x) = f(g(x))$$

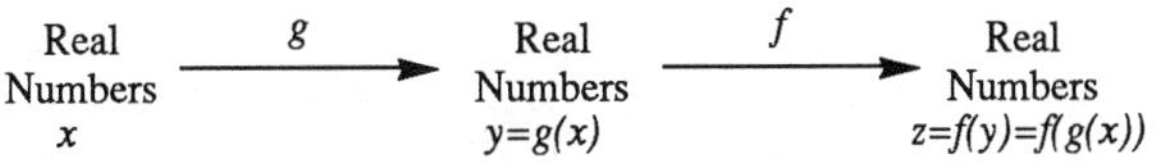

Using the definition of one-to-one function, we can rewrite the conclusion of the given statement as:

B: If x_1 and x_2 are any two distinct real numbers (that is, $x_1 \neq x_2$), it follows that $f \circ g(x_1) \neq f \circ g(x_2)$.

Proof: Let x_1 and x_2 be any two distinct real numbers.

Let's examine in detail each step in the construction of the function $f \circ g$, to see "what happens" to x_1 and x_2.

Since g is a one-to-one function and $x_1 \neq x_2$, it follows that

$$g(x_1) \neq g(x_2).$$

Set $y_1 = g(x_1)$ and $y_2 = g(x_2)$.

Since f is a one-to-one function and $y_1 \neq y_2$, it follows that

$$f(y_1) \neq f(y_2).$$

Thus, if we put these two results together, we can conclude that whenever x_1 and x_2 are any two distinct real numbers,

$$f(g(x_1)) \neq f(g(x_2)).$$

Therefore $f \circ g$ is a one-to-one function.

■

Example 3: If the x- and y-intercepts of a line that does not pass through the point (0,0) have rational coordinates, then the slope of the line is a rational number.

Discussion:

A: Consider a line in the x,y plane. Its x- and y-intercepts are points in the plane with rational coordinates, and neither of them is the point (0,0).

B: The slope of the line described in A is a rational number.

Implicit hypothesis: We know the basic facts about the Cartesian plane; we know how to calculate the slope of a line; we know the coordinates of the x- and y-intercepts of a line; and we know the properties and operations of rational numbers.

The hypothesis mentions two points that belong to the line, namely, its x- and y-intercepts. If we know the coordinates of two points belonging to a line, we can calculate the slope of the line.

Indeed, if $A(x_1,y_1)$ and $B(x_2,y_2)$ are two points on a line, the slope of the line is the number

$$m = \frac{y_2 - y_1}{x_2 - x_1}$$

Proof: By hypothesis, if A is the x-intercept of the line, then $A(p/q, 0)$, with $p \neq 0$ (because A is not the point (0,0)), and $q \neq 0$, and with p and q integers.

By hypothesis, if B is the y-intercept of the line, then $B(0, r/s)$, with $r \neq 0$ (because B is not the point (0,0)), and $s \neq 0$, and with r and s integers.

Therefore

$$m = \frac{\frac{r}{s} - 0}{0 - \frac{p}{q}} = \frac{rq}{sp}$$

with $sp \neq 0$, $rq \neq 0$, and sp and rq integers.

Thus m, the slope of the line, is a rational number.

■

Example 4: The sum of the first n counting numbers is equal to $[n\,(n + 1)]/2$.

Discussion:

A: Consider the sum of the first n counting numbers.

B: Their sum is equal to $[n(n + 1)]/2$; that is:

$$1 + 2 + 3 + \ldots + n = \frac{n(n + 1)}{2}$$

We need to find a pattern to evaluate the sum $1 + 2 + 3 + \ldots + n$.

Proof: Let $S_n = 1 + 2 + 3 + \ldots + (n - 1) + n$.

Addition is a commutative operation. Therefore we can write

$$S_n = n + (n - 1) + \ldots + 2 + 1$$

Let's compare these two ways of writing S_n .

$$S_n = 1 + 2 + \ldots + (n - 1) + n$$

$$S_n = n + (n - 1) + \ldots + 2 + 1$$

If we add these two equalities, we obtain

$$2S_n = (1 + n) + [2 + (n - 1)] + \dots + [(n - 1) + 2] + (n + 1)$$

or

$$2S_n = (n + 1) + (n + 1) + \dots + (n + 1) + (n + 1)$$

or

$$2S_n = n(n + 1)$$

From this last equality we obtain

$$S_n = \frac{n(n + 1)}{2}$$

(This proof is known as Gauss' proof.)

■

Example 5: If a and b are two positive integers, then we can find two integers q and r such that

$$a = bq + r$$

with $0 \le r < b$ and $0 \le q$.

Proof: If a is a multiple of b, then the statement is proved and $r = 0$. Therefore we will assume that a is not a multiple of b.

If we consider the multiples of b, we will find one of them that exceeds a. Call A the collection of all the multiples of b that are larger than a.

By the Well-Ordering Principle, we can pick the smallest element in A. Call it $(q + 1)b$.

Since $(q + 1)b$ is the smallest element in A, the number qb does not belong to A.

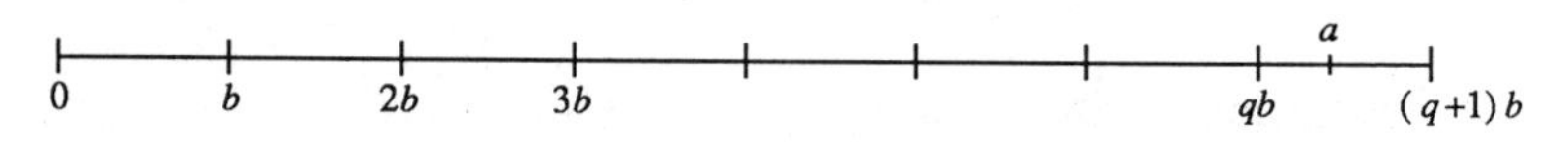

Thus qb is smaller than a. So

$$qb < a < (q + 1)b.$$

If we subtract qb we obtain

$$0 < a - qb < b$$

Set $r = a - qb$.
Then $a = qb + r$.

■

The statement in Example 5 is part of the division algorithm. Later on we will prove that the numbers q and r we just found are the only ones satisfying the required property. (See the section on the uniqueness theorems.)

PROOF BY CONTRAPOSITION

In some cases we cannot use the kind of straightforward arguments we have just seen; that is, we cannot deduce B directly from A, possibly because assuming that A is true does not give us enough information to allow us to prove that B is true. In some cases direct verification of the statement would be too time consuming or impossible. Therefore we must find another starting point.

If it were possible to show that B can be false (by using "not B") when A is true, then the statement we are trying to prove would be false, because B does not follow from A. Therefore, to prove that the statement "A implies B" is true, we can prove that if B is false, then A must be false as well.

This kind of proof is called *proof by contraposition* (or contradiction) because we start from the assumption that B is false and we end up contradicting A, which we knew to be true by hypothesis.

Therefore, when we prove a statement by contraposition, we are really proving that "not B" implies "not A."

Let's consider an example.

Example 6: If x is a rational number and y is an irrational number, their sum $x + y$ is an irrational number.

Discussion: In this case we can set:

A: Consider a rational number x and an irrational number y.

(The fact that the two numbers are called x and y is irrelevant. We could call them a and b, or n and p. We will use x and y to be consistent with the original statement.)

(*Implicit hypothesis:* We can use the properties and operations of real numbers.)

B: Their sum, $x + y$, is an irrational number.

If we wanted to verify the statement directly, we would have to check all possible cases of sums of rational and irrational numbers. Is there a more practical way?

Why should $x + y$ be irrational?

Proof: Let's assume that the opposite is true; that is, we will assume that $x + y$ is a rational number.

Therefore we can write

$$x + y = n/p$$

with $p \neq 0$, and n and p integers.

Since x is rational, $x = a/b$ with $b \neq 0$ and a and b are integers. Thus we have

$$a/b + y = n/p$$

which implies $y = n/p - a/b = (nb - ap)/pb$ with $pb \neq 0$ since $p \neq 0$ and $b \neq 0$.

Since the numbers $nb - ap$ and pb are integers, we can now conclude that y is a rational number.

But this is a contradiction to A because y is known to be irrational. Thus our assumption "$x + y$ is a rational number" is incorrect.

Therefore $x + y$ must be irrational.

■

Sometimes we need to use direct proof and proof by contradiction to construct one proof, as in the following example.

Example 7: Let n be a nonprime counting number. Then n is divisible by a prime number p such that

$$p \leq \sqrt{n}$$

Proof: Since n is nonprime, it is divisible by numbers other than 1 and itself.

Let's consider the collection of all positive divisors of n:

$D = \{$all positive numbers d that divide n, with $1 < d < n\}$.

We know there is at least one number in D, because n is nonprime. So, by the Well-Ordering Principle, D has a smallest number. Call it s.

What do we know about s?

1. s is a divisor of n, since it belongs to D.
2. If c is another divisor of n, then $s \leq c$.
3. The number s is prime.
   Let's prove this claim by contraposition.
   If s is not prime, then $s = ab$ with $1 < a < s$ and $1 < b < s$.
   Since a (and b) divides s and s divides n, it follows that a (and b) divides n.
   Thus a is in the collection D. But this contradicts the fact that s is the smallest element in D. Therefore s must be prime.
4. $s < n$, since it belongs to D.

If we can show that $s^2 \leq n$, then we will have found the number with the required properties.

Since s is a divisor of n, we can write $n = sm$.

The number m divides n as well. Therefore by property 2 of s, $s \leq m$.

Therefore we have

$$s^2 = s\,s \leq s\,m = n.$$

The proof is now complete.

■

Example 8: There are infinitely many prime numbers.

Discussion: We need to analyze the statement to find the hypothesis and the conclusion because they are not clearly distinguishable.

We want to consider the collection of all prime numbers and show that it is infinite. Therefore we can set:

A: Consider the collection of prime numbers. Call it P.
(*Implicit hypothesis:* We can use the properties of prime numbers and the operations and properties of counting numbers, since prime numbers are counting numbers.)
B: P is infinite.

To prove this statement directly, we have to show that we "never run out of prime numbers"; that is, we will never be able to write a complete explicit list of all prime numbers. This approach does not seem reasonable. If the collection is indeed infinite, we are wasting our time. If the collection is finite, it could be very large, and therefore our task would be too time-consuming. We have to find a more practical starting point.

Why should the collection of prime numbers be infinite?

Proof: Let's assume that the conclusion of the statement is not true and that the collection P of prime numbers is indeed finite. Then we can list all its elements.

$$p_1 = 2; p_2 = 3; p_3 = 5; \ldots ; p_n.$$

Using the trichotomy property of real numbers, we can order the n prime numbers we just collected. So

$$p_1 < p_2 < p_3 < \ldots < p_n.$$

Thus p_n is the largest existing prime number.

(*Discussion:* Could we construct another prime number using the ones we listed?

We have seen that the sum of prime numbers is not always a prime number. So we do not know whether

$$p_1 + p_2 + p_3 + \ldots + p_n$$

is prime.

What about the number $p_1 p_2 p_3 \ldots p_n$? This number is not prime because it is divisible by $p_1, p_2, p_3, \ldots,$ and p_n. It is an even number because it is divisible by 2 [remember, $p_1 = 2$]. Prime numbers larger than 2 must be odd. So we are looking for an odd number.)

The number $q = p_1 p_2 p_3 \ldots p_n + 1$ is odd and larger than all the other prime numbers. The number q is not prime, because it is not in the list of prime numbers we have.

Using the result stated in Example 7, we find a prime number t that divides q and

$$t \leq \sqrt{q}.$$

But none of the prime numbers we listed divides q. Therefore another prime number must exist.

We now have a contradiction, because we assumed that we had a complete list of prime numbers.

So our assumption is incorrect, and the collection of prime numbers must be infinite.

■

HOW TO SAY "NO"

Some statements can be proved in more than one way, either by using direct proof or using proof by contraposition. Generally, if we have a choice, we should use the direct technique. Indeed, direct proofs are usually more informative and more intuitively understood.

Moreover, in the proof by contradiction there is a subtle point. We have to deny part B of the statement; that is, we have to construct the statement "not B," which can be a tricky step.

Sometimes it is enough to put the word *not* in B to change it into its opposite, "not B," as in the preceding examples. The statements "$x + y$ is irrational" and "the collection is infinite" become "$x + y$ is not irrational" and "the collection is not infinite."

It is not so easy when B includes words such as: *unique, at least, for one, for all, every, none.* (These expressions are usually called *quantifiers.*)

Let's see how we can change some of these expressions into their opposites.

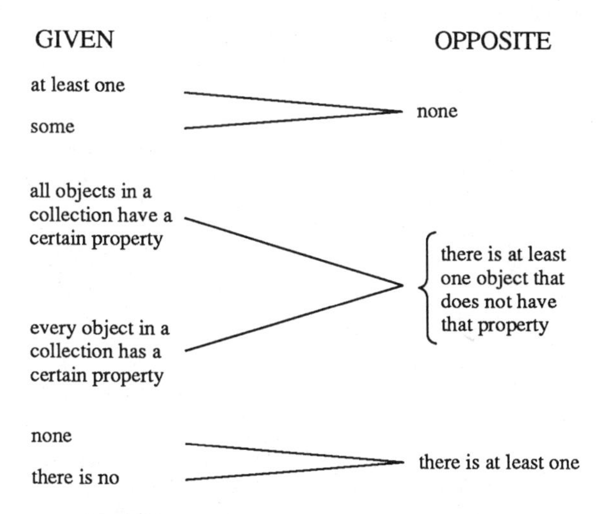

Again, the contraposition method should be used when the assumption that A is true does not give a good starting point, but the assumption that B is false does.

Sometimes the statement we want to prove gives us a hint that it is easiest to use the contraposition method. This method is helpful if B already contains a "not," because if we deny B we get an affirmative sentence.

Example 9: The graphs of the functions $f(x) = -x/6 + 1/4$ and $g(x) = (x - 1)/(x^2 + x - 2)$ have no points in common.

Discussion:

A: Two functions, f and g, are given. Consider their graphs.
(*Implicit hypothesis:* We are familiar with the concepts of function and graph of a function.)

B: The two graphs of f and g do not have any point in common.

We could construct the graphs of both functions. The graph of f is a line, so it is easy to obtain. The graph of g is more complicated. The function g is not defined at $x = -2$ and $x = 1$. Thus its graph has three parts. Therefore it might be easier to use proof by contraposition.

We have to deny the statement "The two graphs do not have any point in common," which is simple to do because the statement is already negative. Its opposite is "The two graphs have at least one point in common."

Proof: Suppose, by way of contradiction, that there is at least one value of the variable x such that

$$f(x) = g(x) \qquad (\#)$$

This equality is not true when $x = -2$ and $x = 1$, because g is not defined for these values. So we will concentrate on $x \neq -2$ and $x \neq 1$, and try to solve the equation

$$-x/6 + 1/4 = (x - 1)/(x^2 + x - 2)$$

$$-x/6 + 1/4 = (x - 1)/(x - 1)(x + 2)$$

We can divide by $x - 1$ because $x \neq 1$, so $x - 1 \neq 0$.

$$-x/6 + 1/4 = 1/(x + 2)$$

We can multiply by $x + 2$ because $x \neq -2$, so $x + 2 \neq 0$.

So we obtain $2x^2 + x + 6 = 0$. This equation has no real solutions, because its discriminant is negative ($\Delta = 1^2 - 4(6)(2) = -47$). Thus the equation (#) cannot be true. Therefore the two graphs have no points in common.

■

Example 10: There is no positive real number smaller than 1 that is smaller than its square.

Discussion:

A: Consider the collection of positive real numbers smaller than 1.
(*Implicit hypothesis:* We are familiar with the definition, properties, and operations of real numbers.)

B: None of the numbers described in A is smaller than its square.

Let's rewrite these statements in a more manageable way:

A: Let x be any real number, with $0 < x < 1$.
B: x is not smaller than x^2.

We will prove this statement in two ways and compare the two proofs:

1. Direct proof

We know that $x < 1$.

We can multiply both sides of this inequality by the positive number x and obtain $x^2 < x$. Thus, if $0 < x < 1$ the statement is true, because x is not smaller than x^2.

2. Proof by contraposition

We need to construct the statement "not B."

"not B": There is at least one number x, with $0 < x < 1$, that is smaller than x^2.

We can use the inequality

$$x < x^2 \tag{\#}$$

as the starting point of our proof.

Since we can use the properties and operations of real numbers, we can rewrite the previous inequality as

$$x - x^2 < 0.$$

The expression on the left-hand side of the inequality can be factored as

$$x(1 - x) < 0.$$

The product of two real numbers is negative only if one of the numbers is positive and the other is negative.

By hypothesis $x > 0$.

Thus we must conclude that $1 - x < 0$. But $1 - x < 0$ implies that $1 < x$. This contradicts the fact that x is a number smaller than 1. So "not B" cannot be true. Therefore B must be true.

■

In Example 10 the direct proof is shorter and simpler than the proof by contradiction. Therefore, even if we are considering a statement formulated in a negative way, we might decide to use the direct proof.

EXERCISES

Given the following statements, construct their opposites.

1. The function f is defined for all real numbers.
2. Given the two numbers x and y, there is a rational number z such that $x + z = y$.
3. The function f has the property that if x and y are two distinct real numbers, then $f(x) \neq f(y)$.
4. The equation $P(x) = 0$ has only one solution. (Assume we know that the equation has at least one solution.)
5. All nonzero numbers have nonzero opposites.
6. For every number $n > 0$, there is a corresponding number $M_n > 0$ such that $f(x) > n$ for all numbers x with $x > M_n$. (To understand this statement better, try to get a graph in the Cartesian plane.)
7. Every number satisfying the equation $P(x) = Q(x)$ is such that $|x| < 5$.
8. The equation $P(x) = 0$ has only one solution. (Check Exercise 4.)
9. The function f is continuous at the point c if for every $\varepsilon > 0$ there is a $\delta>$ 0 such that if $|x - c| < \delta$, then

$$|f(x) - f(c)| < \varepsilon.$$

10. For every real number x, the number $f(x)$ is rational.

Using the techniques of this section, prove the following statements.

1. Let f and g be two nondecreasing functions such that the function $f \circ g$ exists. Then this function is nondecreasing.

   (A function h is called nondecreasing if for every two real numbers x and y such that $x \leq y$, $h(x) \leq h(y)$. The function $f \circ g$ is defined as $f \circ g(x) = f(g(x))$.)

2. If x is a rational nonzero number and y is an irrational number, then the number xy is irrational.
3. Let n be a number with three or more digits. If the two-digit number made by n's last two digits is divisible by 4, n is divisible by 4.
4. If $(a + b)^2 = a^2 + b^2$ for all real numbers b, then a must be zero.
5. Let n be a counting number. If the number $2^n - 1$ is a prime number, then n is a prime number as well.
6. Every four-digit palindrome number is divisible by 11. (A palindrome number reads the same forward or backward.)
7. Let f be a nondecreasing function for all real numbers. Then

$$\frac{f(c) - f(x)}{c - x} \geq 0$$

for all $x \neq c$. (See Exercise 1.)
8. Prove the following statement in two ways: directly and using the contraposition method.

   The function $f(x) = mx + b$, with $m \neq 0$, is a one-to-one function. (For the definition of one-to-one function, see Example 3 in this section.)
9. Let f and g be two real-valued functions, defined for all real numbers.

   If f and g are onto, so is $f \circ g$.

   (A function h is onto if for every real number y there is at least one real number x such that $h(x) = y$.)

Read the following proofs and make sure you understand them. Then outline the proofs, listing explicitly all the most important steps. Fill in details that might have been skipped (such as: write the statement in the form "if A*, then* B*," indicate which technique has been used, fill in algebraic steps, and so on).*

10. *Euclid's algorithm for the greatest common divisor of two numbers*

    Let a and b be two positive integers with $a > b$.

    Divide a by b and write

$$a = bq_1 + r_1$$

with $q_1 \geq 0$ and $0 \leq r_1 < b$.

Then divide b by r_1, obtaining

$$b = r_1 q_2 + r_2$$

with $q_2 \geq 0$ and $0 \leq r_2 < r_1$.

Continuing, we can divide r_1 by r_2 and obtain

$$r_1 = r_2 q_3 + r_3$$

with $q_3 \geq 0$ and $0 \leq r_3 < r_2$.

Continue this process while $r_k \neq 0$.

Then the greatest common divisor of a and b is the last nonzero remainder.

The greatest common divisor of a and b, usually denoted as (a,b) or GCD(a,b), is a positive integer number d such that:

1. d divides both a and b
2. any other divisor of a and b divides d

Proof: If we use the process described in the statement, we obtain

$$a = bq_1 + r_1$$

$$b = r_1 q_2 + r_2$$

$$r_1 = r_2 q_3 + r_3$$

....

....

$$r_{n-3} = r_{n-2} q_{n-1} + r_{n-1}$$

$$r_{n-2} = r_{n-1} q_n + r_n$$

$$r_{n-1} = r_n q_{n+1} + 0$$

The process will take at most b steps because $b > r_1 > r_2 > \ldots \geq 0$. The last equality implies that $r_n = \text{GCD}(r_{n-1}, r_n)$. (Explain why.)

Since

$$r_{n-2} = r_{n-1} q_n + r_n$$

$$= r_n q_{n+1} q_n + r_n = r_n t_1 \text{ with } t_1 > 0.$$

it follows that r_n divides r_{n-2} and r_{n-1}. So r_n is a common divisor of r_{n-2} and r_{n-1}.

If d is another positive integer divisor of r_{n-2} and r_{n-1}, then d will divide r_n.(Check this claim.) Therefore $r_n = \text{GCD}(r_{n-1}, r_{n-2})$.

Similarly, working backward through all the equalities of the algorithm, we obtain that $r_n = \text{GCD}(r_{n-2}, r_{n-3})$, ... , $r_n = \text{GCD}(a,b)$.

■

11. If $d = \text{GCD}(a,b)$, then $d = sa + tb$ for some integers s and t. Therefore, if a and b are relatively prime (that is, $\text{GCD}(a,b) = 1$), then

$$1 = sa + tb \qquad \text{for some integers } s \text{ and } t.$$

Proof: Using the steps of the Euclidean algorithm, we obtain

$$r_1 = a - bq_1$$

$$r_2 = b - r_1 q_2 = b - (a - bq_1)q_2 = as_2 + bt_2$$

$$r_3 = r_1 - r_2 q_3 = (a - bq_1) - (as_2 + bt_2)q_3 = as_3 + bt_3$$

Proceeding in this way, in at most b steps we will be able to write

$$r_n = sa + tb$$

The statement is therefore proved.

■

12. Let p be a prime number. If p divides the product ab, then p divides either a or b.

Proof: If p does not divide a, then $1 = \text{GCD}(a,p)$. (Explain why.) Then

$$1 = sa + pt$$

for some integers s and p.

Thus

$$b = b(sa + pt)$$

$$= abs + bpt = kps + bpt$$

$$= p(ks + bt)$$

This implies that p divides b.

■

13. Let p be a prime number.

Then $\sqrt{p}$ is an irrational number.

Proof: Let's assume that $\sqrt{p}$ is a rational number, that is,

$$\sqrt{p} = \frac{n}{q}$$

with $n \neq 0$, $q \neq 0$, n and q integers, and n/q in lowest terms.

Therefore

$$p = \frac{n^2}{q^2}$$

So $n^2 = pq^2$.

Since n^2 is a multiple of p and p is a prime number, n must be a multiple of p. Therefore we can write $n = pk$ for some positive integer k. This implies

$$p^2k^2 = pq^2$$

or

$$pk^2 = q^2$$

Since q^2 is a multiple of p, and p is a prime number, q must be a multiple of p. Thus $q = pm$ for some positive integer m. Therefore

$$\frac{n}{q} = \frac{pk}{pm} = \frac{k}{m}$$

This contradicts the fact that n/q is already in lowest terms, and proves that the square root of a prime number is irrational.

■

Special Kinds of Theorems

There are certain kinds of theorems whose proofs follow standard patterns. In this chapter we list the *most important* of these special theorems.

"IF AND ONLY IF" OR EQUIVALENCE THEOREMS

This kind of statement is rather common and it is very useful in mathematics. If we can show that "A is true if and only if B is true" (or that "A is false if and only if B is false"), we are proving that A and B are equivalent statements, because the truth of one automatically implies the truth of the other.

Therefore we could use either A or B to describe A certain property, depending on which of the two statements better fits our approach at that point.

The statement "A if and only if B" means that "A is a necessary and sufficient condition for B" and "B is a necessary and sufficient condition for A."

Thus to prove that this statement is true we have to show that:

1. If A, then B
(A is sufficient for B; B is necessary for A)

and

2. If B, then A
(B is sufficient for A; A is necessary for B)

Therefore we need to split the proof of an "if and only if" statement into two parts. We can use any of the techniques we have seen to prove each part.

Statements 1 and 2 are the *converses* of each other. The converse of a statement of the form "if A, then B" is obtained by interchanging the hypothesis and the conclusion, which reads "if B, then A."

Example 1: A nonzero real number is positive if and only if its reciprocal is positive.

Proof:
A: A real number a is positive.
B: The reciprocal of a, a^{-1}, is positive.

1. If A, then B
(The fact that the number a is positive is sufficient to imply that its reciprocal is positive.)

Since $a \cdot a^{-1} = 1$, $a \cdot a^{-1}$ is a positive number.
The product of two numbers is positive only if they are either both positive or both negative. Since the number a is positive by assumption, it follows that a^{-1} must be positive.

2. If B, then A
(The fact that the number a is positive is necessary for its reciprocal to be positive.)

By definition of reciprocal

$$a \cdot a^{-1} = 1.$$

Since the product of a and a^{-1} is a positive number, a and a^{-1} must be either both positive or both negative. The number a^{-1} is positive by hypothesis, so a is positive.

■

Example 2: Let n be a counting number.
n is odd if and only if n^2 is odd.

Proof:
A = n is odd.
B = n^2 is odd.

1. A implies B.
(The fact that n is odd is sufficient to imply that n^2 is odd.)

Since n is odd we can write $n = 2p + 1$. Therefore

$$n^2 = (2p + 1)^2 = 4p^2 + 4p + 1 = 4(p^2 + p) + 1$$

Since $4(p^2 + p)$ is an even number, $4(p^2 + p) + 1$ is odd.
Thus n^2 is odd.

2. B implies A.
   (The fact that n is odd is necessary to imply that n^2 is odd.)

(*Discussion:* If n^2 is odd we can write $n^2 = 2t + 1$.
Therefore $n = \sqrt{n^2} = \sqrt{2t + 1}$
This step does not give us any useful information. REMEMBER:

$$\sqrt{2t + 1} \neq 2\sqrt{t} + 1.$$

So we need to look for another starting point. Let's try to use proof by contraposition.
If we know that n^2 is odd, why should we conclude that n is odd?)

Let's assume that n is not odd. Then it must be even: $n = 2t$.
Thus $n^2 = 4t^2$ is even, since 4 times any integer is even.
This is a contradiction because n^2 is odd by hypothesis. Therefore n must be odd.

■

As we can see from Example 2, the proof of an equivalence theorem might require the use of direct proof for one implication and proof by contraposition for the other implication.

Some theorems list three or more statements and claim that they are all equivalent. The proofs of this kind of theorem are rather flexible (that is, they can be set up in different ways), as long as we show that each statement implies each of the others (either directly or through some intermediate steps) and that each statement is implied by each of the others. In this way we prove that each statement is sufficient and necessary for all the others.

Let's assume that we want to prove that four statements A, B, C, and D are equivalent. There are many ways of proceeding. Let's look at four of them. We will look in detail at the first chain of implications and just give the diagrams of some of the other possible chains. It will be up to you to check that by proving the implications indicated by the arrows of each diagram, we would indeed show that the four statements A, B, C, and D are equivalent.

1.

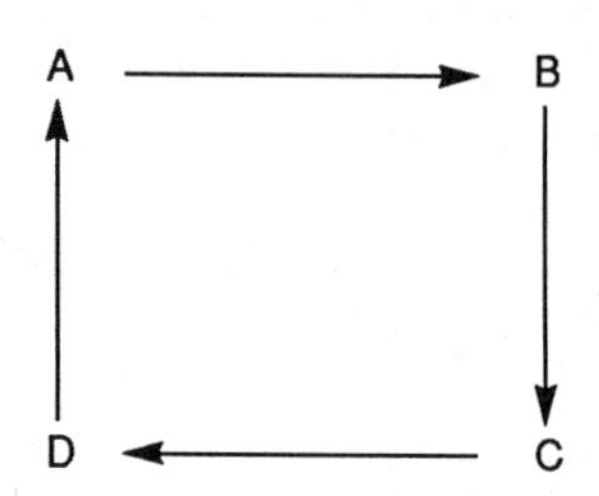

We use this diagram to show that we proved the following:

(i) If A, then B
(ii) If B, then C
(iii) If C, then D
(iv) If D, then A

(The order in which the four statements have been proved is not relevant, unless we want to use one of the implications in the proof of another. We can use only results that we have already proved.)

We can see that if these four implications are true, then A, for example, implies B, C, and D. Indeed:

a. A implies B (directly proved).
b. A implies B, and B implies C; so A implies C.
c. A implies C, and C implies D; so A implies D.

A follows from B, C, and D. Indeed, D implies A (proved directly). C implies D, that implies A; therefore C implies A. B implies C, that implies A; so B implies A.

In the same way, we can check that B, C, and D imply all the other statements and are implied by all of them.

2.

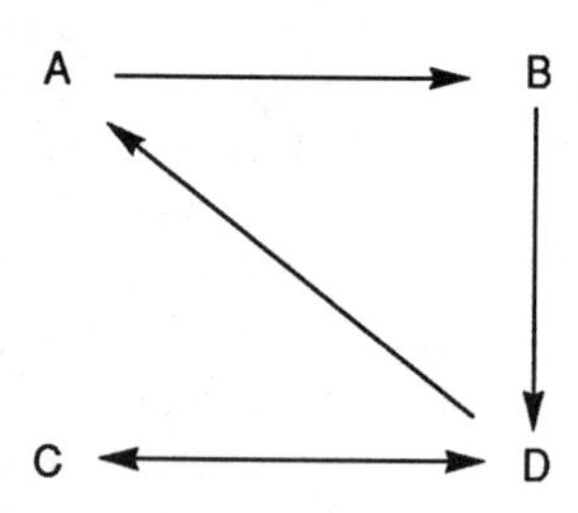

3.

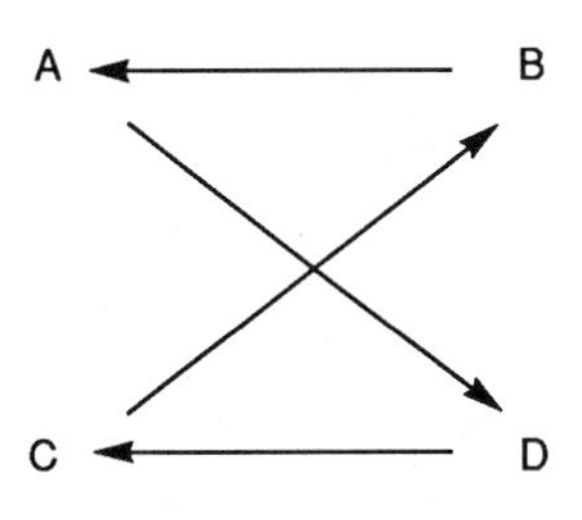

4.

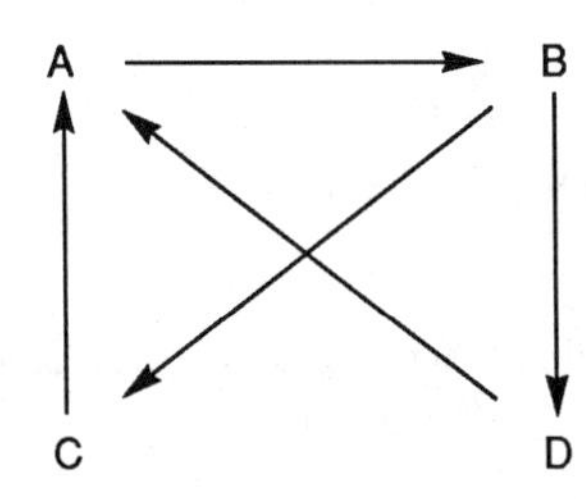

Depending on our priorities, we can choose either the chain of proofs that involves the implications that are easier to prove or the one that gives more detailed information. Therefore, there is no unique way of proving that three or more statements are equivalent.

Before proceeding any further we would like to emphasize the usefulness of knowing that two or more statements are equivalent. Indeed, their equivalence enables us to use the statement that is more suitable for our purpose.

It is valuable to keep in mind that the following statements are always equivalent:

1. B if A
2. If A, then B
3. A implies B
4. B is a necessary condition for A
5. A is a sufficient condition for B
6. A only if B
7. "not B" implies "not A" (contrapositive)

The contrapositive, the converse, and the inverse of a statement are not equivalent, in general.

Let's summarize their definitions. Given the statement "if A, then B," we can construct three statements:

- *contrapositive* "not B" implies "not A"
  (equivalent to the given statement)

- *converse* "if B, then A"
  (not equivalent to the given statement)

- *inverse* "not A" implies "not B"
  (not equivalent to the given statement)

We will now look at an example to clarify these concepts. Consider the true statement:

"If x is a rational number, then x^2 is a rational number."

Its *contrapositive*, which is equivalent to the given statement, is the true statement:

"If x^2 is not a rational number, then x is not a rational number."

Its *converse* is the statement:

"If x^2 is a rational number, then x is a rational number," which in this example is a false statement. (Consider, as an example, $x^2 = 3$.)

Its *inverse* is the statement:

"If x is not a rational number, then x^2 is not a rational number," which happens to be false as well. (Consider $x = 2^{1/2}$.)

Let's now consider some more examples.

Example 3: Let a and b be two distinct real numbers. Then the following statements are equivalent:

(i) b is larger than a.

(ii) Their average, $\frac{a+b}{2}$, is larger than a.

(iii) Their average, $\frac{a+b}{2}$, is smaller than b.

Discussion: By hypothesis a and b are real numbers, so we can use all the properties of real numbers, such as the order (or trichotomy) property, since the given statements deal with the comparison of numbers.

The order property states that, given any two distinct real numbers, it is always possible to compare them and to decide which is the smaller and which is the larger.

Since we are dealing with the average of two numbers, we can recall the fact that the average of two numbers is always larger than the smaller of them and smaller than the larger of them. This is the kind of result one might want to outline in a lemma and prove separately.

Lemma. Let a and b be two distinct real numbers. Then their average is always larger than the smaller number and smaller than the larger number.

Proof of the lemma: Since a and b are not equal, let's assume that $a < b$.

$$a=\frac{1}{2}a+\frac{1}{2}a<\frac{1}{2}a+\frac{1}{2}b<\frac{1}{2}b+\frac{1}{2}b=b$$

So

$$a<\frac{a+b}{2}<b$$

■

Now that we have gathered some important information, we will prove the theorem by using the following chain of steps:

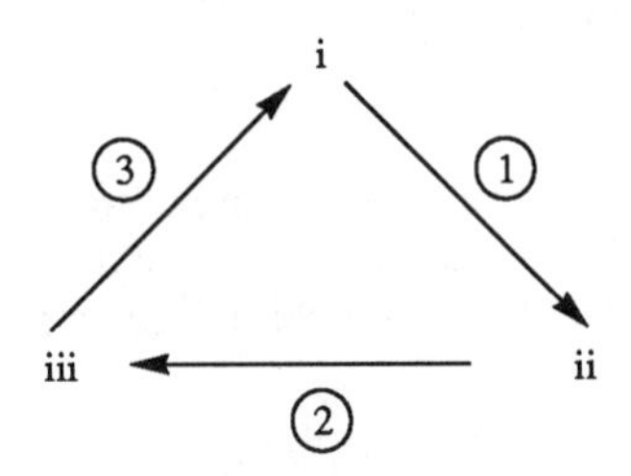

Proof:

1. If (i), then (ii)
(That is: If b is larger than a, then their average is larger than a.)
This result follows from the lemma.

2. If (ii), then (iii)
By the lemma we know that the average of two numbers is larger than one of them and smaller than the other.
Therefore, if $\frac{a+b}{2}$ is larger than a, it must be smaller than b.

3. If (iii), then (i)
We are assuming that $\frac{a+b}{2} < b$.

We want to be able to compare a and b. Therefore we will try to simplify the given inequality

$$\frac{a+b}{2}<b$$

Thus

$$a+b<2b$$

So

$$(a+b)-b<2b-b$$

and

$$a<b$$

This proves that (i) is true.
The proof is now complete.

■

Often we need to prove that two or more correct definitions of the same object are equivalent. The existence of different definitions is usually generated by different approaches that emphasize some properties and points of view over others.

Example 4: The following definitions are equivalent:

(i) A triangle is an isosceles triangle if it has two equal sides.
(ii) A triangle is an isosceles triangle if it has two equal angles.

Proof:

1. (i) implies (ii)
We have to show that if a triangle has two equal sides, then it will have two equal angles.

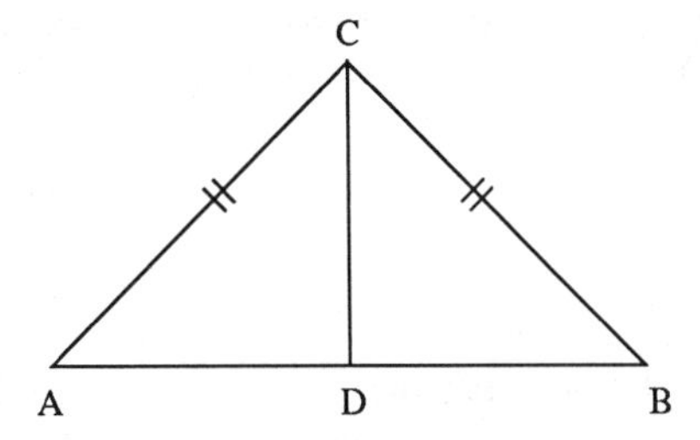

Consider the two triangles *ADC* and *CDB* obtained by considering *CD*, perpendicular to the base *AB*.

The angles made at the intersection of *CD* and *AB* are equal, because they are right angles.

The two triangles have two equal sides: *CD*, because it is a common side, and *AC* is equal to *CB* by hypothesis. Thus *AD* and *DB* are equal (we can use the Pythagorean theorem). This implies that the two triangles *ADC* and *CDB* are congruent and the two angles at the vertices *A* and *B* are equal.

2. (ii) implies (i)
We have to show that if a triangle has two equal angles, then it has two equal sides.

Let's assume that the angles at the vertices *A* and *B* of the triangle *ABC* are equal. Let's again consider the triangles *ADC* and *CDB* obtained by considering *CD* perpendicular to *AB*.

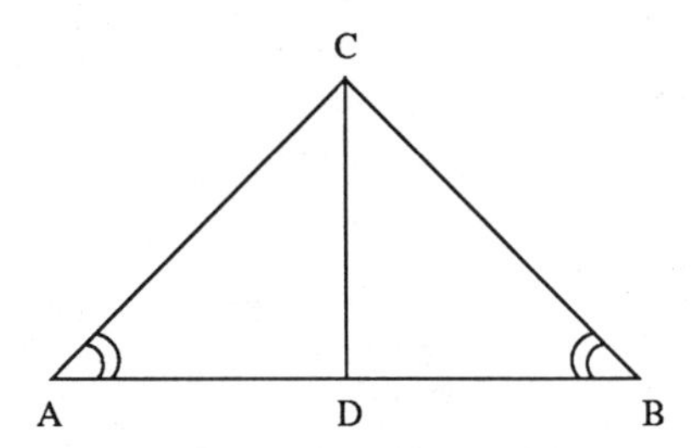

The two triangles are similar. Indeed, the angles made at the intersection of *CD* and *AB* are equal (because they are right angles), and the angles at the vertices *A* and *B* of the triangle are equal by hypothesis. Moreover, the two similar triangles *ADC* and *CDB* have the side *CD* in common; therefore they are congruent. In particular, the sides *AC* and *CB* are equal.

■

EXERCISES

Given the following statements, write their (a) contrapositives, (b) converses, and (c) inverses.

1. If x is an integer divisible by 6, then x is divisible by 2.
2. If a quadrilateral is not a parallelogram, its diagonals do not bisect.
3. If $P(x) = a_n x^n + a_{n-1} x^{n-1} + \ldots + a_1 x + a_0$ and $Q(x) = b_n x^n + b_{n-1} x^{n-1} + \ldots + b_1 x + b_0$ are equal for all real numbers, then $a_i = b_i$ for all i, with $0 \leq i \leq n$.

Prove the following statements.

1. A function f is nonincreasing ([hat is: If x and y are such that $x \leq y$, then $f(x) \geq f(y)$] if and only if $(f(c) - f(x))/(x - c) \leq 0$ for all c and x in the domain of f with $x \neq c$.

2. Let r and s be two counting numbers. The following statements are equivalent:

   (i) $r > s$.
   (ii) $a^s < a^r$ for all real numbers $a > 1$.
   (iii) $a^r < a^s$ for all real positive numbers $a < 1$.

3. Let a and b be two distinct real numbers. Then the following statements are equivalent:

   (i) b is larger than a ;
   (ii) their average, $\frac{a+b}{2}$, is larger than a;
   (iii) their average, $\frac{a+b}{2}$, is smaller than b.

   Prove this statement by showing that (1) statements (i) and (ii) are equivalent, and (2) statements (i) and (iii) are equivalent.

4. Let x and y be two distinct, negative real numbers. Then the following statements are equivalent:

   (i) $x < y$
   (ii) $|x| > |y|$
   (iii) $x^2 > y^2$

5. Consider the two systems of linear equations

S_1

$$a_1x + b_1y = c_1$$
$$a_2x + b_2y = c_2$$

S_2

$$a_1x + b_1y = c_1$$
$$(a_1 + ba_2)x + (b_1 + bb_2)y = c_1 + bc_2$$

with $a_1, a_2, b_1, b_2, b, c_1, c_2$ real numbers and $b \neq 0$.

The pair of values (x_0,y_0) is a solution of S_1 if and only if it is a solution of S_2.

USE OF COUNTEREXAMPLES

An example can be very useful when trying to make a point or explain the result obtained in a proof, but it cannot be used as a proof.

Let's see what could happen if we used and accepted examples as proofs. Someone could claim that if a and b are any two real numbers, then

$$(a + b)^2 = a^2 + b^2 \qquad (\ddagger)$$

"How can you have any doubt about it?" that person could say. "There are so many examples of pairs of numbers satisfying this equality!"

Take $a = 0$ and $b = 1$, for example.

$$(a + b)^2 = (0 + 1)^2 = 1$$

$$a^2 + b^2 = 0^2 + 1^2 = 1$$

So, it is true that $(a + b)^2 = a^2 + b^2$.

Take, for example, $a = 0$ and $b = -1$; $a = 0$ and $b = 2$; $a = -4$ and $b = 0$; and so on. For all these pairs the equality $(a + b)^2 = a^2 + b^2$ holds true. But if you look at all the pairs of numbers used to show that the equality works, you can see that in all of them at least one number is equal to zero.

The equality is claimed to hold true for *all possible* pairs of numbers, not only for some special pairs. What happens if we take $a = 1$ and $b = 2$?

$$(a+b)^2 = 3^2 = 9$$

$$a^2 + b^2 = 1^2 + 2^2 = 1 + 4 = 5$$

Therefore the original statement (‡) is false, in spite of all the examples. We have found a *counterexample.*

So we can see that examples do not work as proofs that a certain property and/or equality holds true for *all* the objects in a collection (unless the collection is finite so that it is possible to check that the property and/or equality holds true for each single element in the collection), because examples deal with special cases and do not verify that the given statement is true in general. If we can find at least one example for which the statement is not true, we say that we have found a counterexample and that the counterexample is a proof that the given statement is false in at least one instance. Therefore it is false in general.

Consider the statement "Every real number a has a reciprocal." We can think of thousands of numbers that do have a reciprocal. But the existence of only *one* number with no reciprocal (the number zero does not have a reciprocal) makes the statement "*Every* real number a has a reciprocal" false. What is true is that "*Every nonzero* real number has a nonzero reciprocal."

Sometimes the existence of a counterexample can help us understand why a certain statement is not true and whether some restrictions in the hypothesis would change it into a true statement.

For example, the statement "The equality

$$(a+b)^2 = a^2 + b^2$$

holds for all pairs of real numbers a and b in which at least one of the two numbers is zero" is a true statement. (Prove it.)

The discovery of a counterexample can save the time and effort spent looking for a proof that is not going to work, but sometimes even counterexamples are difficult to find.

There is no sure way of knowing when to look for a counterexample. If your attempts to construct a proof have failed, despite having used a systematic approach, then look for a counterexample. This search might be difficult. But if it is successful, you can stop looking for a proof, because you just proved that the statement is false. If it is unsuccessful, then you might be finding examples in which the statement holds. The examples you found can give you an insight into why the statement is true, which will help you construct the proof.

Example 1: For all real numbers $x > 0$, $x^3 > x^2$.

Discussion: You might want to get a better handle on this problem statement by graphing the functions x^3 and x^2.

We can break this statement into

A: x is a positive real number (so we can use properties and operations of real numbers)

B: $x^3 > x^2$

Proof : Let's look for a counterexample. If we take $x = 0.5$, then $x^3 = 0.125$ and $x^2 = 0.25$. Therefore, in this case, $x^3 < x^2$. So the statement is false.

■

Example 2: If two sides of a triangle are equal, then the triangle is equilateral. (•)

Discussion: There is no reason why all the triangles that have two sides of the same length must have the third side equal in size to the first two. But someone might not be persuaded by this kind of reasoning.

Proof: Can we find a counterexample?

The triangle

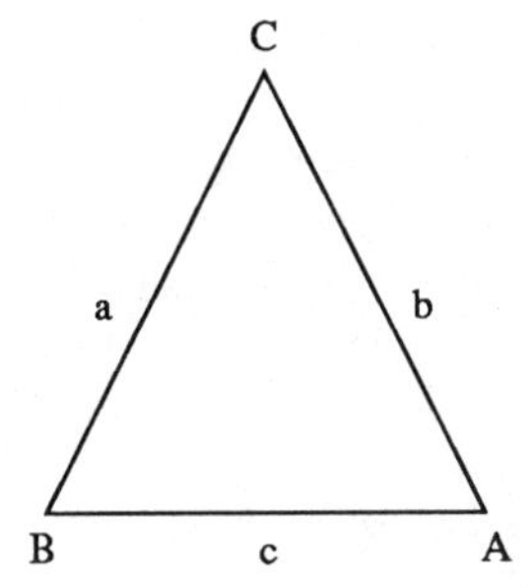

with a = b = 5
and c = 3

has two sides of the same length, but it is not equilateral. This counterexample shows that the statement (•) is false.

■

Example 3: If an integer is a multiple of 10 and of 15, then it is a multiple of 150.

Proof: The statement is false. Just consider the least common multiple of

10 and 15, namely 30. This number is a multiple of 10 and of 15, but it is not a multiple of 150.

■

EXERCISES

Use counterexamples to show the following statements to be false.

1. Let f be an increasing function and g be a decreasing function. (That is: If $x \leq y$, it follows that $f(x) \leq f(y)$ and $g(x) \geq g(y)$.)
   The function $f + g$ defined as $(f + g)(x) = f(x) + g(x)$ is constant.
2. If t is an angle in the first quadrant, then $2 \sin t = \sin 2t$.
3. Consider the polynomial $P(x) = -x^2 + 2x - 3/4$. The variable $y = P(x)$ is always negative.
4. The reciprocal of a real number $x \geq 1$ is a number y, with $0 < y < 1$.
5. The number $3n + 2$ is prime for all counting numbers n.
6. Let f, g, and h be three functions defined for all real numbers. If $f \circ g = f \circ h$, then $g = h$.

Discuss the truth of the following statements; that is, prove those that are true and give a counterexample for those that are false.

7. The sum of any five consecutive integers is divisible by 5.
8. If $f(x) = x^2$ and $g(x) = x^4$, then $f(x) \leq g(x)$ for all real numbers $x \geq 0$.
9. The sum of four consecutive counting numbers is divisible by 4.
10. Let f and g be two odd functions, defined for all real numbers. Their sum, $f + g$, defined in Exercise 1, is an even function defined for all real numbers.
    (A function h is called odd if $h(-x) = -h(x)$ for all x. A function h is called even if $h(-x) = h(x)$ for all x.)
11. Let f and g be two odd functions, defined for all real numbers. Their quotient f/g defined as

$$\frac{f}{g}(x) = \frac{f(x)}{g(x)}$$

is an even function defined for all real numbers x.

MATHEMATICAL INDUCTION

We use this kind of proof when we need to show that a certain statement holds true for a large collection of natural numbers and therefore direct verification is difficult, if not impossible.

We cannot simply check that a statement is true for *some* numbers and then generalize the result to the whole collection (this procedure is called *incomplete induction*, and its use in the case we are considering is incorrect). Again, examples are not proofs.

Consider the following claim: The inequality

$$n^2 \leq 5n!$$

is true for all counting numbers $n \geq 3$.

How many numbers should we check? Is the statement true since the inequality holds true for $n = 3, 4, 5, 6, \ldots, 30$? We cannot check directly all counting numbers $n \geq 3$. Therefore we must look for another way to prove this kind of statements.

The mathematical induction (or *complete induction*) technique consists of the following steps:

1. Prove that the statement is true for the smallest number that can be used in the given statement.
2. Assume that the statement is true for an arbitrary natural number n (inductive hypothesis).
3. Using hypotheses 1 and 2 show that the statement is true for the next number, namely $n + 1$ (deductive step).

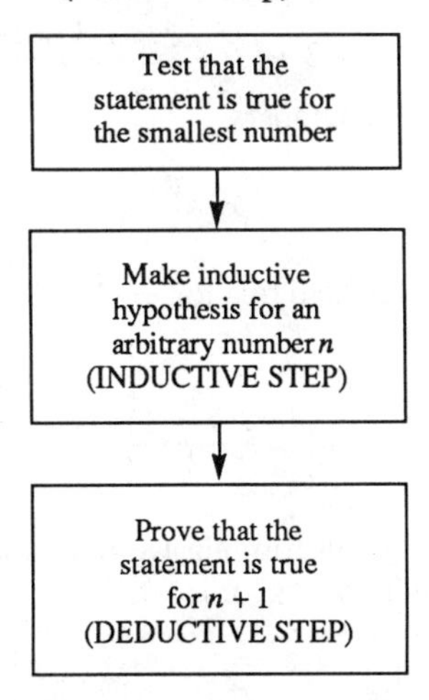

As you can see, the construction of the proof by induction is quite peculiar. We start by checking that the given statement is true in a special case. We know that we cannot stop here because examples are not proofs. Then we "trust" the statement temporarily, and we check its strength by using deductive reasoning to see if the result can be extended one step further. If the statement is true, it will pass this last test.

At this point it is possible to see how one can conclude that the statement is true for all the numbers in a given collection. Indeed we proved that the statement is true for the smallest number in the collection and that if the statement is true for a number it will be true for the next one.

Since the statement is true at least for the smallest number, it will be true for the next number, and then for the next, and for the next, and so on for all the other numbers.

The fact that this extension process can go on indefinitely would require more explanation, which is beyond the purpose of this book. (Intuitive understanding is sufficient at this stage.) Indeed, mathematical induction is founded on a very important result in number theory: The Principle of Mathematical Induction. This Principle can be stated as follows.

If $P(n)$ represents a statement relative to a positive integer n, then $P(n)$ is true for all $n \geq t$ (where t is a fixed positive integer) provided that:

1. $P(t)$ is true
2. If $P(t), P(t + 1), \ldots, P(n)$ are true, then $P(n + 1)$ is true.

Example 1: Prove by induction that the sum of the first k natural numbers $1, 2, \ldots, k$ is equal to $[k(k + 1)]/2$; that is

$$\underbrace{1 + 2 + 3 + \ldots + k}_{k \text{ numbers}} = \frac{k(k + 1)}{2}$$

Proof:

1. Does the given formula hold true for $k = 1$, the smallest number that can be used?

$$1 = \frac{1(1 + 1)}{2}$$

So by using the given formula we obtain a true statement. This means that the formula works for $k = 1$.

2. Let's assume that the formula works when we add the first n numbers ($k = n$). Thus

$$\underbrace{1 + 2 + 3 + \ldots + n}_{n \text{ numbers}} = \frac{n(n+1)}{2}$$

3. We want to prove that the formula holds true when n is replaced by $n + 1$. Thus we have to show that

$$\underbrace{1 + 2 + 3 + \ldots + n + (n+1)}_{(n+1) \text{ numbers}} = \underbrace{\frac{(n+1)[(n+1)+1]}{2}}_{\text{formula}}$$

or equivalently

$$1 + 2 + 3 + \ldots + n + (n+1) = \frac{(n+1)(n+2)}{2}$$

To do this we want to use the formula we are assuming to be true for the first n numbers; that is,

$$1 + 2 + 3 + \ldots + n = \frac{n(n+1)}{2}$$

$1 + 2 + 3 + \ldots + n + (n+1)$	associative property for the addition of numbers
$= [1 + 2 + 3 + \ldots + n] + (n+1)$	use the formula for the addition of the first n numbers
$= \dfrac{n(n+1)}{2} + (n+1)$	common denominator
$= \dfrac{n(n+1) + 2(n+1)}{2}$	common factor
$= \dfrac{(n+1)(n+2)}{2}$	

Thus

$$1 + 2 + 3 + \ldots + n + (n+1) = \frac{(n+1)(n+2)}{2}$$

Therefore the formula given in the statement works for any natural number by induction.

■

Example 2: The sum of the first k odd numbers is equal to k^2.

$$1 + 3 + 5 + \ldots + (2k - 1) = k^2$$

Proof: Check $k = 1$. This means that we are considering only one odd number, namely 1. In this case we have

$$1 = 1^2$$

and the equality is therefore true.

Let's assume that

$$1 + 3 + 5 + \ldots + (2n - 1) = n^2$$

We want to prove that the equation is true for $n + 1$. Therefore we need to check the equality

$$1 + 3 + 5 + \ldots + (2n - 1) + (2(n + 1) - 1) = (n + 1)^2$$

Using the associative property of addition we can write

$$1 + 3 + 5 + \ldots + (2n - 1) + (2(n + 1) - 1)$$

$$= \{1 + 3 + 5 + \ldots + (2n - 1)\} + (2n + 1) \quad \text{by inductive hypothesis}$$

$$= n^2 + (2n + 1) = (n + 1)^2$$

The equality is therefore correct.

■

Example 3: If $n > 1$ is a counting number, then either n is a prime number, or a product of prime numbers.

Proof: The statement is true for the smallest number we can use, which is 2.

Let's now assume that the statement is true for $2, 3, \ldots, n - 1$. Is it true for n?

If n is a prime number, the statement is trivially true.

If n is not a prime number, by definition n has a positive divisor d such that $d \neq 1$ and $d \neq n$.

So $n = dm$ with $m \neq 1$ and $m \neq n$ and $m \geq 0$.

Since both d and m are positive numbers larger than or equal to 2 and smaller than n, by inductive hypothesis they are either prime or products of prime numbers.

If they are both prime numbers, the statement is proved.

If at least one of them is not prime, we can replace it with its prime factors. So n will be a product of prime factors in any case.

■

We would like to mention that it is possible to prove that the Principle of Mathematical Induction is equivalent to the Well-Ordering Axiom.

EXERCISES

1. For all positive integers k,

$$1 + 2 + 2^2 + 2^3 + \ldots + 2^{k-1} = 2^k - 1$$

2. Prove that 8 divides $9^k - 1$ for all $k \geq 1$.
3. Check that for all $k \geq 1$

$$2 + 4 + 6 + \ldots + 2k = k^2 + k$$

4. Let $a > 1$ be a fixed number. For all integer numbers $k \geq 3$

$$(1 + a)^k > 1 + ka^2$$

5. Check that for all counting numbers $k \geq 1$

$$1/2 + (1/2)^2 + (1/2)^3 + \ldots + (1/2)^k = \frac{1 - \left(\frac{1}{2}\right)^{k+1}}{1 - \frac{1}{2}} - 1$$

6. The inequality $k^2 \leq 5k!$ is true for all counting numbers $k \geq 3$.
7. Let n be an odd counting number. Then $n^2 - 1$ is divisible by 4.

EXISTENCE THEOREMS

Existence theorems claim that at least one object having certain properties exists. This kind of theorem can be proved in two ways:

1. If it is possible, find an algorithm (a rule) to explicitly construct at least one object satisfying the requirements stated in the theorem.

2. Sometimes, especially in higher-level mathematics, the explicit construction of the object is not possible; therefore we must obtain a general existence statement for the kind of object in consideration, without ever finding an actual example of the given type of object.

Example 1: Given rational numbers a and b, with $a < b$, there exists a rational number c such that $a < c < b$.

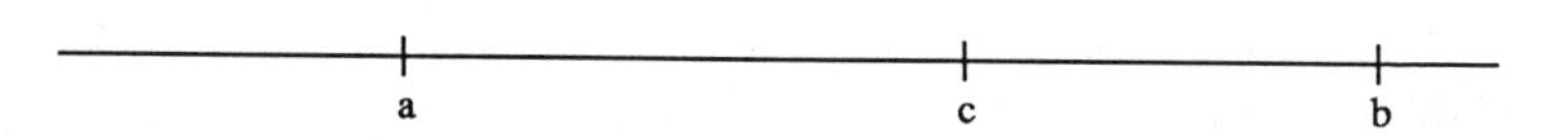

Discussion: We can rewrite the statement as: If a and b are two rational numbers, with $a < b$, then there exists a rational number c such that $a < c < b$.

Let's recall the definition of rational numbers: A number is said to be rational if it can be written as the quotient (ratio) x/y of two integers x and y with $y \neq 0$. We prefer to have x/y written in lowest terms.

Proof: We are now looking for a number c that is "between" a and b, and which is rational. Let's take their average

$$c = \frac{a+b}{2}$$

Can we prove that c is still a rational number?

Indeed, if $a = \frac{m}{n}$ and $b = \frac{p}{q}$ with $n \neq 0$, $q \neq 0$, we have

$$c = \frac{1}{2}\left(\frac{m}{n} + \frac{p}{q}\right) = \frac{1}{2}\frac{mq + np}{nq}$$

$$c = \frac{mq + np}{2nq}$$

The number $mq + np$ is an integer because m, q, n, p are integers. The number $2nq$ is an integer, and it is different from zero because $n \neq 0$ and

$q \neq 0$. Therefore c is a rational number.

Now we have to show that c is larger than a and smaller than b. We have already shown that the average of two numbers is smaller than one and larger than the other, so we can use that result, without going through the proof again.

This proves that c is the number with the required properties.

■

Example 2: A line passes through the points (0,2) and (2,6).

Proof: Let's reformulate the statement as: If (0,2) and (2,6) are points in the plane, then we can find a line passing through them.

1. Using method 1 we are going to write explicitly the equation of the line with the required properties.

   The point-slope equation of a line is $y - y_0 = m(x - x_0)$ where m is the slope (m = [(change in y)]/[(change in x)] = 4/2 = 2) and (x_0, y_0) are the coordinates of any point on the line.

   Therefore the line we are looking for has the equation

$$y - 2 = 2(x - 0)$$

   or

$$y = 2x + 2$$

■

[We can check that the points (0,2) and (2,6) are on the line by showing that their coordinates satisfy the equation we have found.

Check: a. (0,2) is on the line. Use $y = 2$ and $x = 0$. $2 = 2(0) + 2$
b. (2,6) is on the line. Use $y = 6$ and $x = 2$. $6 = 2(2) + 2$]

2. Using method 2 we only want to prove that we can find a line through (0,2) and (2,6). There is a postulate (from geometry) that states that given any two distinct points in the plane there is a unique straight line joining them. Therefore there is a line through (0,2) and (2,6).

■

Example 3: The polynomial $P(x) = x^4 + x^3 + x^2 + x - 1$ has a real zero between 0 and 1.

Discussion: If we want to use method 1, we need to find the zeros of the polynomial $P(x)$ by solving the fourth-degree equation

$$x^4 + x^3 + x^2 + x - 1 = 0$$

The process of solving a fourth-degree equation uses cumbersome formulas, even if it is not difficult. Since we are not asked to find the precise value of the zeros of the polynomial, but only to prove that at least one of them is located between 0 and 1, we could look for a different approach.

Graph the polynomial $P(x)$ for $0 \le x \le 1$.

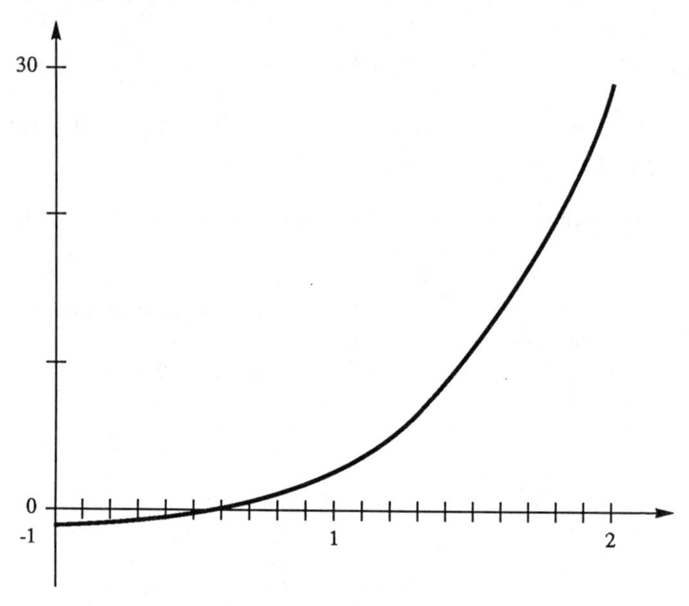

Proof: Since $P(0)$ is a negative number, and $P(1)$ is a positive number, the graph of $P(x)$ intersects the x-axis between $x = 0$ and $x = 1$. Indeed, polynomials are continuous functions, so we can use the intermediate value theorem. Therefore there is a value x_0 of the x-variable such that $y = P(x_0) = 0$. The statement is now proved.

■

EXERCISES

Prove the following statements.

1. There exists a function whose domain consists of all the real numbers and whose range is the interval $[0,1]$.

2. There is a counting number n such that $2^n + 7^n$ is a prime number.
3. Let a be an irrational number. Then there exists an irrational number b such that ab is an integer.
4. There is a second-degree polynomial P such that $P(0) = -1$ and $P(-1) = 2$.
5. There exist two rational numbers a and b such that a^b is a positive integer, and b^a is a negative integer.
6. If $P(x) = a_n x^n + a_{n-1} x^{n-1} + ... + a_0$ is a polynomial of degree n, with n odd, and $a_n \neq 0$, then the equation $P(x) = 0$ has at least one real solution.

UNIQUENESS THEOREMS

This kind of theorem states that an object (whose existence has already been proved or is assumed to be true) is unique with respect to certain properties; that is, it is the only object having the required properties.

In order to prove the uniqueness of an object in a given collection, we have to show that no other object in that collection satisfies the properties listed in the statement. Direct checking is usually impossible because the collection can be very large (or infinite).

Therefore we have to find a different approach. What would happen if the object with the required properties were not unique? (Proof by contraposition.) To assume that the special object is not unique means to consider the possibility that there is at least one more object with the same properties. If the uniqueness statement is true, the two elements will have to coincide, showing that there is only *one* object with the desired properties.

Example 1: The number, usually indicated by 1, such that for all real numbers a

$$a \cdot 1 = 1 \cdot a = a$$

is unique.

Proof: Let t be a number with the property that

$$a \cdot t = t \cdot a = a$$

for all real numbers a (even for $a = 1$).

(We cannot use the symbol 1 for it, because as far as we know t could be different from 1.)

The number 1 leaves every other number unchanged when multiplied by it. So

$$t \cdot 1 = t$$

The number t leaves every number unchanged when multiplied by it. So

$$t \cdot 1 = 1$$

Therefore

$$t = t \cdot 1 = 1$$

Thus $t = 1$. Therefore it is true that there is a unique number with the described property.

■

Very often existence and uniqueness theorems are combined in a statement of the form: There exists a unique . . .

To prove that a statement of this kind is true we have to show that

1. The object exists and
2. The object is unique.

Example 2: If $f(x) = x^3$, for all real numbers x, there exists a unique function g such that

$$g \circ f(x) = g(f(x)) = x \text{ and } f \circ g(x) = f(g(x)) = x$$

for all real numbers x. g is called the inverse function of f with respect to composition.

Proof: We have to prove that a function with the listed properties exists and that it is unique.

1. g exists.

Since the function f is described by an algebraic expression, let's look for an algebraic expression for g.

The function g has to be such that

$$f(g(x)) = x$$

Therefore

$$(g(x))^3 = x$$

or

$$g(x) = x^{1/3} = \sqrt[3]{x}$$

Let's check whether g really has the required properties.

$$g \circ f(x) = g(x^3) = \sqrt[3]{x^3} = x$$

$$f \circ g(x) = f(\sqrt[3]{x}) = (\sqrt[3]{x})^3 = x$$

2. g is unique.

In this case we establish the uniqueness of g in two ways:

a. We can say that g is unique by the way it is defined, or
b. Let h be a function such that

$$h \circ f(x) = x \qquad f \circ h(x) = x$$

for all real numbers x.

We want to compare h and g.

By definition two functions are equal if they agree on all the elements in their domain (they must have the same domain).

g and h are defined for all real numbers.

At this point we need to compare these two functions by using their properties:

$$g(x) = g(f \circ h(x)) \quad \text{associative property of composition}$$

$$= g \circ f(h(x)) = h(x)$$

Therefore $g = h$.

So the inverse of f is unique.

■

EXERCISES

Prove the following statements.

1. The polynomial $p(x) = x - b$ with $b \in \mathbb{R}$, has a unique zero.
2. There exists a unique angle θ with $0 \leq \theta \leq \pi$ such that $\cos\theta = \theta$.
3. The equation $x^3 - b = 0$, with b a real number, has a unique solution.
4. There exists a unique second-degree polynomial P such that $P(0) = -1$, $P(1) = 3$, and $P(-1) = 2$.
5. We proved that if n is an integer number larger than 1, then n is either prime or a product of prime numbers. Thus we can write

$$n = p_1 p_2 \ldots p_k$$

where the p_j are prime numbers and $p_1 \leq p_2 \leq \ldots \leq p_k$.
Show that this factorization of n is unique.

Outline the proofs of the following statements, showing all the details.

1. Let f be a function defined for all real numbers. If f is one-to-one and onto, its inverse exists and it is unique.
(By definition, a function is one-to-one if whenever x_1 and x_2 are two distinct real numbers, then $f(x_1) \neq f(x_2)$.
A function f is onto if for every y there is at least one x such that $f(x) = y$.)

Proof: We need to find a function g defined for all real numbers such that

$$g \circ f(x) = g(f(x)) = x$$

$$f \circ g(y) = f(g(y)) = y$$

for all real numbers x and y.

Let y be any real number. Define $g(y) = x$, where x is the unique number such that $f(x) = y$. (Why do we know that such a number x exists and that it is unique for all real numbers y?)

We claim that the function g so defined is the inverse of the function f.

Check: Let x_0 be a real number such that $f(x_0) = y_0$. Thus

$$g \circ f(x_0) = g(f(x_0)) = g(y_0) = x_0$$

On the other hand, if $g(y_0) = x_0$, it follows that $f(x_0) = y_0$. So

$$f \circ g(y_0) = f(g(y_0)) = f(x_0) = y_0$$

Assume that there is another function h which is the inverse of f. Therefore

$$h \circ f(x) = x \qquad f \circ h(y) = y$$

for all real numbers x and y. Thus

$$f \circ h(y) = y = f \circ g(y)$$

for all real numbers y. This implies that

$$f \circ h(y) = f \circ g(y)$$

Since f is a one-to-one function, we obtain

$$h(y) = g(y)$$

for all real numbers y. Therefore $h = g$.

■

2. **Division Theorem.** Let a and b be two integers such that $a \geq 0$ and $b > 0$. Then there exist two unique integers q and r, with $q \geq 0$ and $0 \leq r < b$ such that $a = bq + r$.

   (We have already proved the existence of numbers with the required properties. We want you to read a different proof of the existence statement. This proof uses mathematical induction.)

Proof: We will prove the existence of numbers like the ones described in the conclusion by considering different cases.

1. If $b = 1$, we can consider $q = a$ and $r = 0$. Therefore we will consider $b > 1$.
2. If $a = 0$, just take $q = 0$ and $r = 0$. So the statement is true.
3. The statement is clearly true for all numbers $a < b$. Take $q = 0$ and $r = a$.
4. Assume that $a \geq b$. We will prove by induction that in this case there ex-

ist two integers q and r, with $q \geq 0$ and $0 \leq r < b$, such that $a = bq + r$.

If $a = b$, the statement is true: Take $q = 1$ and $r = 0$.

Assume that the statement is true for a number $n > b$. Then consider $a = n + 1$.

$$a = n + 1 = (bq_1 + r_1) + 1$$

$$= bq_1 + (r_1 + 1)$$

Since $0 \leq r_1 < b$, $1 \leq r_1 + 1 \leq b$.

If $r_1 + 1 < b$, then just set $q_1 = q$ and $r_1 + 1 = r$.

If $r_1 + 1 = b$, we can write

$$a = bq_1 + b = bq + r$$

where $q = q_1 + 1$ and $r = 0$.

We need to prove the uniqueness part of the statement. Let's assume that there are two pairs of integers q and r, q' and r' such that

$$a = bq + r = bq' + r'$$

$q \geq 0$ and $0 \leq r < b$, $q' \geq 0$ and $0 \leq r' < b$.

Because of the trichotomy property of integers, we can consider two cases: $r' \leq r$ or $r' \geq r$. Let's assume that $r' \geq r$.

Thus $b(q - q') = r' - r \geq 0$.

Since $r' > r' - r$, $b > r' - r$. Therefore $b > b(q - q') \geq 0$. So if we divide by b, we obtain

$$1 > q - q' \geq 0$$

This implies $q - q' = 0$, or $q = q'$.

If $r' \leq r$, just consider $b(q' - q) = r - r'$, and proceed using the same logic.

The theorem is now completely proved.

■

EQUALITY OF SETS

A set is a well-defined collection of objects. The objects that belong to a set are called the elements of the set.

To show that two sets, A and B, are equal (that is, have the same elements) we need to prove two statements:

1. $A \subset B$ (A is contained in B)
2. $B \subset A$ (B is contained in A)

A set A is contained in a set $B (B \supseteq A)$ if every element of A is an element of B. Therefore, if we can show that B contains A and is contained in A at the same time, we will have shown that all the elements of B belong to A and all the elements of A belong to B. This means that A and B have the same elements, and therefore they are equal.

Example 1: Let $A = \{n \in \mathbb{Z} \mid \text{the remainder of } n/2 \text{ is zero}\}$ and $B = \{\text{all the integer multiples of } 2\}$. Prove that $A = B$.

Proof:

1. $A \subset B$

   Let x be a generic element of A (that is, x is any number satisfying the conditions needed to belong to the collection A.) We have to show that x belongs to B as well.

   Since x is in A we can write $x = 2q + 0$; that is, $x = 2q$. This means that x is a multiple of 2. Therefore x is an element of B.

2. $B \subset A$

   Let x be any number in B. We have to check whether x belongs to A. Since x is in B, it is a multiple of 2. So we can write x as $x = 2q + 0$. Since the remainder of the division of x by 2 is 0, x belongs to A.

■

In some cases it is easier to compare two or more sets after making their descriptions more explicit.

Example 2: Let $A = \{x \in \mathbb{R} \text{ such that } |x/2 - 1| < 5\} = \{x \in \mathbb{R} \mid |x/2 - 1| < 5\}$ and $B = \{x \in \mathbb{R} \mid x \text{ is a number between the roots of the equation } x^2 - 4x - 96 = 0\}$. Prove that the two sets are equal.

Proof:
Let's simplify the descriptions of the two sets.
By definition of absolute value, the inequality $|x/2 - 1| < 5$ is equivalent to the inequalities

$$-5 < x/2 - 1 < 5$$

Since $-5 < x/2 - 1$ and $x/2 - 1 < 5$, we obtain $-4 < x/2$ and $x/2 < 6$. Therefore x is a number such that $-8 < x < 12$. Thus we can rewrite

$$A = \{x \in \mathbb{R} \mid -8 < x < 12\}$$

The solutions of the equation $x^2 - 4x - 96 = 0$ are the numbers -8 and 12, since $(-8)^2 - 4(-8) - 96 = 0$ and $(12)^2 - 4(12) - 96 = 0$. Therefore

$$B = \{x \in \mathbb{R} \mid -8 < x < 12\}$$

At this point it is evident that the two sets are equal.

■

In order to construct more interesting examples, we will consider two operations between sets.
Given two sets A and B, the set

$$A \cup B = \{x \mid x \in A \text{ or } x \in B\}$$

is called the **union** of A and B.
The set

$$A \cap B = \{x \mid x \in A \text{ and } x \in B\}$$

is called the **intersection** of A and B.

Example 3: If A, B, and C are any three sets, then

$$A \cap (B \cup C) = (A \cap B) \cup (A \cap C)$$

(distributive property of the intersection with respect to the union).

Proof:
1. $A \cap (B \cup C) \subset (A \cap B) \cup (A \cap C)$

Let $x \in A \cap (B \cup C)$. We want to show that $x \in (A \cap B) \cup (A \cap C)$.

Since $x \in A \cap (B \cup C)$, then $x \in A$ and $x \in (B \cup C)$.

Therefore $x \in A$ and either $x \in B$ or $x \in C$ (or both).

So we can consider two cases: either $x \in A$ and $x \in B$, or $x \in A$ and $x \in C$.

Thus either $x \in A \cap B$ or $x \in A \cap C$, that is, $x \in (A \cap B) \cup (A \cap C)$.

2. $(A \cap B) \cup (A \cap C) \subset A \cap (B \cup C)$

   Let $x \in (A \cap B) \cup (A \cap C)$. We want to show that $x \in A \cap (B \cup C)$.

   Since $x \in (A \cap B) \cup (A \cap C)$, then either $x \in (A \cap B)$ or $x \in (A \cap C)$.

   So either $x \in A$ and $x \in B$, or $x \in A$ and $x \in C$.

   (Therefore x is definitely in A, and either $x \in B$ or $x \in C$.)

   Thus $x \in A$ and $x \in B \cup C$. So we can conclude that $x \in A \cap (B \cup C)$.

■

Example 4: If $A = \{x \in \mathbb{Z} \mid x \text{ is a multiple of } 5\}$ and $B = \{x \in \mathbb{Z} \mid x \text{ is a multiple of } 7\}$, then

$$A \cap B = \{x \in \mathbb{Z} \mid x \text{ is a multiple of } 35\}$$

Proof:

1. Let $x \in A \cap B$. Then $x \in A$ and $x \in B$. This implies that x is a multiple of 5 and of 7. Therefore $x = 5n$ and $x = 7m$.
   If we combine these equalities we obtain $5n = 7m$.
   Since 5 and 7 are prime numbers, the product $5n$ is divisible by 7 only if n is divisible by 7.
   So $n = 7k$ for some integer k. Thus we obtain $x = 5n = 35k$.
   Therefore x is a multiple of 35.
2. Let x be a multiple of 35. Therefore $x = 35t$ for some integer t.
   Thus x is divisible by 5 (so $x \in A$), and x is divisible by 7 (so $x \in B$).
   Therefore $x \in A \cap B$.

■

"Venn diagrams" (John Venn was a British logician [1834–1923]) are commonly used to illustrate properties and operations between unspecified sets. Usually the sets are denoted by discs, which are labeled and placed inside a bigger rectangle that represents the universal set U.

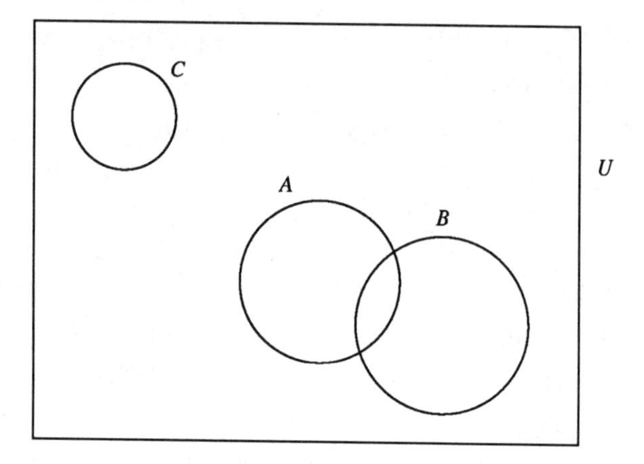

The shaded area in the following diagram represents the set $A \cap B$, the intersection of A and B.

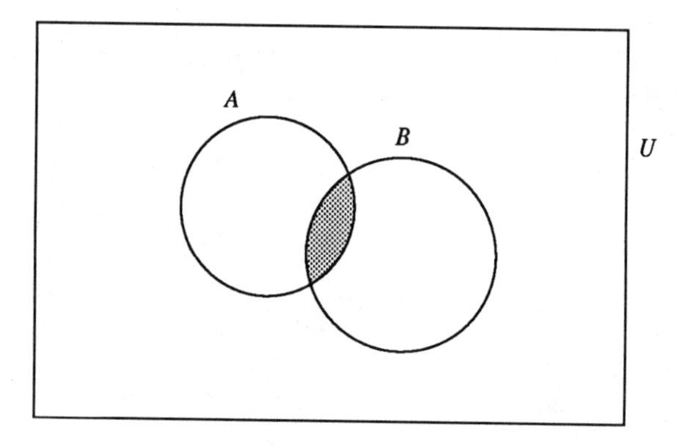

The shaded area in the following diagram represents the set $A \cup B$, the union of A and B.

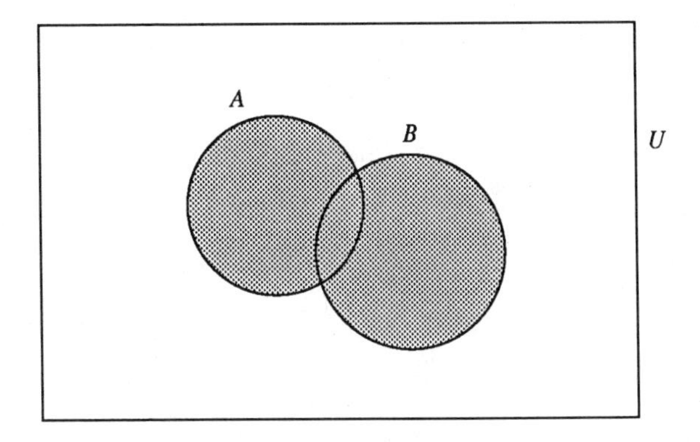

Let's check the equality $A \cap (B \cup C) = (A \cap B) \cup (A \cap C)$.
We will start by constructing the set $A \cap (B \cup C)$.

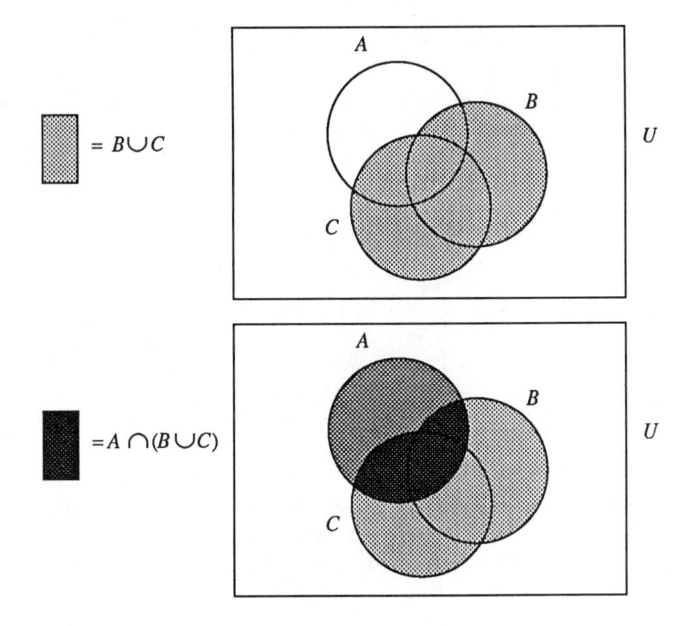

Let's now construct the set $(A \cap B) \cup (A \cap C)$.

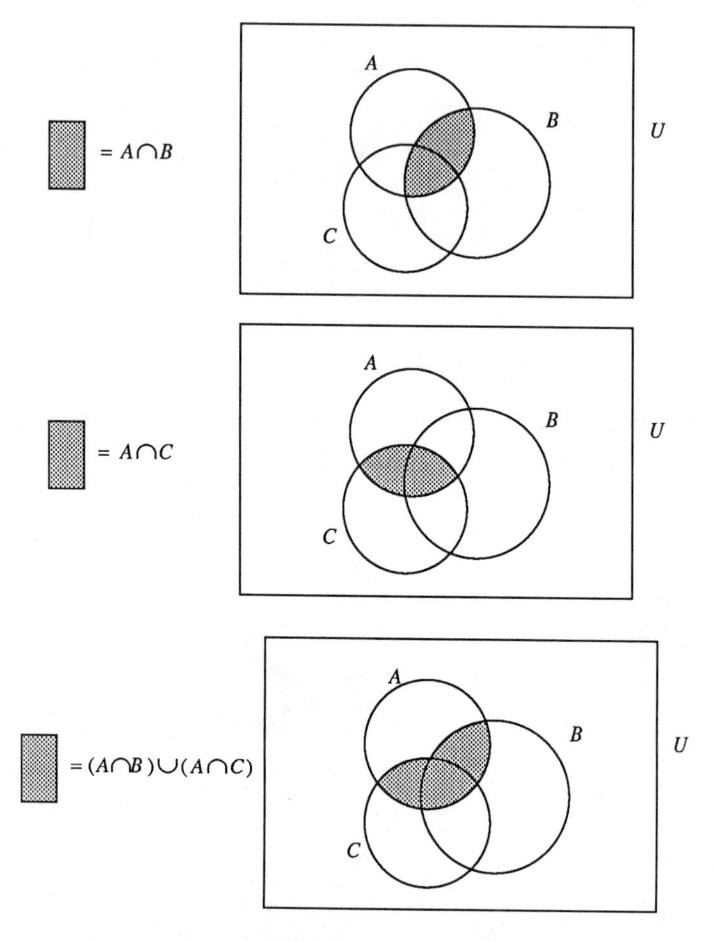

So the two sets obtained using the Venn diagrams are equal.

The use of Venn diagrams does not give a proof of an equality in the general case, but it offers a good illustration of the set operations performed.

The **complement** of a set A is the set of all elements that belong to the universal set U but do not belong to A. The complement of A is usually denoted either by A' or $C(A)$.

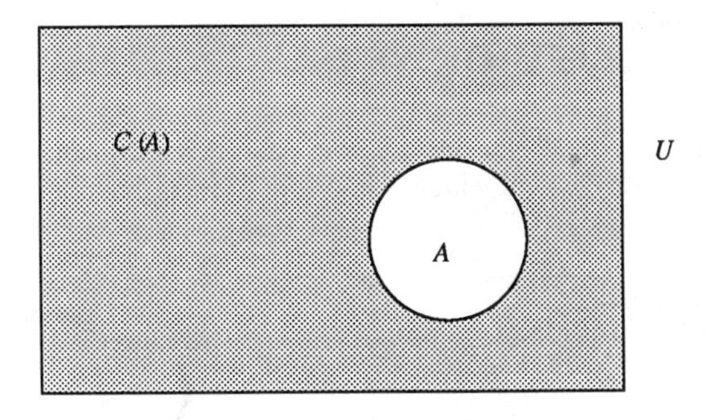

Example 5: Let $A \subset U$ and $B \subset U$. Then

$$(A \cap B)' = A' \cup B'$$

This is one of De Morgan's laws. (The proof of the other law, namely $(A \cup B)' = A' \cap B'$, is left as an exercise.)

Proof:

1. $(A \cap B)' \subset A' \cup B'$

   Let $x \in (A \cap B)'$. This implies that $x \notin A \cap B$. Therefore either $x \notin A$ or $x \notin B$ (because if x belonged to both A and B, then it would belong to their intersection). Thus either $x \in A'$ or $x \in B'$. This implies that $x \in A' \cup B'$.

2. $A' \cup B' \subset (A \cap B)'$

   Let $x \in A' \cup B'$. Then either $x \in A'$ or $x \in B'$. Therefore either $x \notin A$ or $x \notin B$. This implies that x is not a common element for A and B; that is, $x \notin A \cap B$. So we can conclude that $x \in (A \cap B)'$.

■

Sometimes the two inclusions can be proved at the same time. We could have proved the statement in Example 5 as follows:

$x \in (A \cap B)'$ if and only if $x \notin A \cap B$ if and only if $x \notin A$ or $x \notin B$ if and only if $x \in A'$ or $x \in B'$ if and only if $x \in A' \cup B'$.

NON-EQUALITY OF SETS

If we want to show that two sets are not equal, it is sufficient to prove that there is at least one element which is in one of the sets, but not in the other.

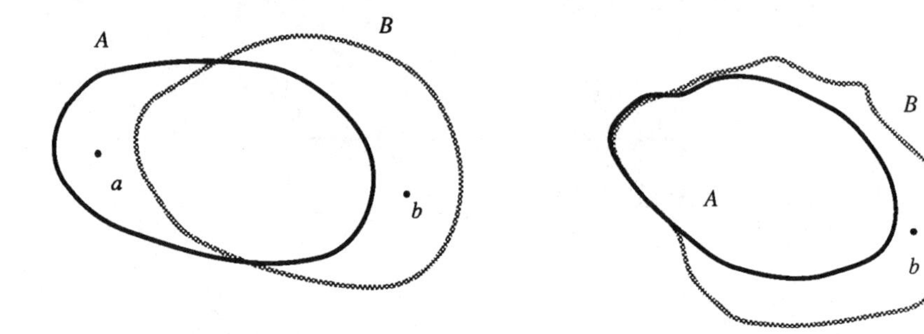

Example 6: Let A = {all the odd counting numbers larger than 1} and B = {all the prime numbers larger than 2}. Are these two sets equal?

Proof: The answer is: no.

We have already seen that all prime numbers larger than 2 are odd. Therefore $B \subset A$.

Are all odd numbers prime numbers? No, 9 is odd, but it is not prime. Therefore the sets are not equal.

■

Example 7: Let C = {all continuous functions on the interval (-1,1)} and D = {all differentiable functions on the interval (-1,1)}. Are these two sets equal?

Proof: The answer is: no.

All differentiable functions are continuous, but not all continuous functions are differentiable. Consider $f(x) = |x|$.

■

EXERCISES

1. Prove that for any three sets A, B, and C

$$(A \cup B) \cup (A \cup C) = A \cup (B \cup C)$$

2. Prove that the sets A = {all the integer multiples of 2 and 3} and B = {all the integer multiples of 6} are equal.

3. Prove the second of De Morgan's laws:

$$(A \cup B)' = A' \cap B'$$

(August De Morgan, 1806–1871, was one of the first to use letters and symbols in abstract mathematics.)

4. Prove that for any three sets A, B, and C

$$(A \cap B) \cap C = A \cap (B \cap C)$$

(associative property of intersection)

5. Prove or disprove the following statement:

   The sets A = {all the integer multiples of 16 and 36} and B = {all the integer multiples of 576} are equal.

6. Prove or disprove the following equalities, with A, B, and C subsets of a universal set U.

   a. $A \cup (B \cap C) = (A \cup B) \cap C$

   b. $(A \cap B \cap C)' = A' \cup B' \cup C'$

7. Prove that the sets

$$A = \{(x,y) \mid y = x^2 - 1 \text{ with } x \in \mathbb{R} \text{ and } y \in \mathbb{R}\}$$

and

$$B = \{(x,y) \mid y = \frac{x^4 - 1}{x^2 + 1} \text{ with } x \in \mathbb{R} \text{ and } y \in \mathbb{R}\}$$

are equal.

8. A set $A \subset \mathbb{R}$ is convex if, whenever x and y are elements of A, the number $tx + (1 - t)y$ is an element of A for all values of t with $0 \leq t \leq 1$.
   The set $\{z \mid z = tx + (1 - t)y \text{ for } 0 \leq t \leq 1\}$ is called the line segment joining x and y.
   Empty sets (sets with zero elements) and sets with one element are assumed to be convex.

Outline the following proof:

The intersection of two or more convex sets is a convex set.

Proof: We will prove the statement using mathematical induction on the number of sets.

Let A_1 and A_2 be two convex sets. If $A_1 \cap A_2$ is either empty or it contains only one element, then it is convex.

Let's assume that $A_1 \cap A_2$ has at least two distinct elements, x and y.

Thus $x, y \in A_1$ and $x, y \in A_2$. Since A_1 and A_2 are convex, the line segment joining x and y is contained in A_1 and in A_2. Therefore it is contained in $A_1 \cap A_2$. So $A_1 \cap A_2$ is convex.

Let's assume that if $A_1, A_2, \ldots, A_n$ are convex sets, then $A_1 \cap A_2 \cap \ldots \cap A_n$ is a convex set.

Let's show that if $A_1, A_2, \ldots, A_{n+1}$ are convex sets, then $A_1 \cap A_2 \cap \ldots \cap A_{n+1}$ is a convex set.

We can use the associative property of intersection to write

$$A_1 \cap A_2 \cap \ldots \cap A_{n+1} = (A_1 \cap A_2 \cap \ldots \cap A_n) \cap A_{n+1}$$

The set $A_1 \cap A_2 \cap \ldots \cap A_n$ is convex by inductive hypothesis.

So $A_1 \cap A_2 \cap \ldots \cap A_{n+1}$ is convex because it is the intersection of two convex sets.

EQUALITY OF NUMBERS

There are at least three ways to prove that two numbers are equal. We can do this by showing that

1. $a \le b$ and $a \ge b$, or
2. $a - b = 0$, or
3. If we know that the numbers are different from zero, we can prove that $a/b = 1$.

It is preferable to use the second and third ways if we can set up algebraic expressions involving a and b. The first way is more useful when we have to compare numbers through the examination of their definitions.

Example 1: If a and b are two positive integers, then their least common multiple is equal to the quotient between ab and the greatest common divisor of a and b; that is,

$$lcm(a,b) = \frac{ab}{GCD(a,b)}$$

Proof: Let $d = \text{GCD}(a,b)$ and $L = lcm(a,b)$
By definition of GCD(a,b), we can write

$$a = dn \qquad \text{and} \qquad b = dp$$

with n and p positive, relatively prime integers. Therefore

$$M = \frac{ab}{d} = \frac{(dn)(dp)}{d} = dnp$$

Clearly M is a multiple of both a and b (indeed, $M = pa$ and $M = nb$). Therefore M is a common multiple of a and b.

Is M the least common multiple of a and b; that is, is $L = M$? By definition of least common multiple we know that $L \le M$. Therefore we only need to show that $L \ge M$.

We will do so by proving that

$$\frac{L}{M} \ge 1$$

By definition L is a multiple of a and b. Thus

$$L = at \qquad \text{and} \qquad L = bs$$

with t and s positive integers. Therefore

$$L = dnt = dps$$

So

$$nt = ps$$

This implies that n must divide ps. Since p and n are relatively prime, it follows that n divides s. Thus $t = pk$ for some integer $k \geq 1$. So

$$L = dnpk$$

Thus

$$\frac{L}{M} = k \geq 1$$

Since $L \leq M$ and $L \geq M$, we can conclude that $L = M$.

■

Example 2: Let $f(x) = x/(x^2 + 1)$, and let x and y be two real numbers larger than 1. Prove that $f(x) = f(y)$ implies that $x = y$.

(We say that f is one-to-one on $(1, +\infty)$.)

Proof: Our starting point is the equality $f(x) = f(y)$. This implies

$$\frac{x}{x^2 + 1} = \frac{y}{y^2 + 1}$$

We can now multiply by $(x^2 + 1)(y^2 + 1)$ because the numbers $(x^2 + 1)$ and $(y^2 + 1)$ are both $\neq 0$. Therefore we obtain

$$xy^2 + x = x^2y + y$$

or

$$(x - y)(1 - xy) = 0$$

Thus either $x - y = 0$ or $1 - xy = 0$.

The first equality implies $x = y$. The second equality implies $xy = 1$. This is impossible because $x > 1$ and $y > 1$.

Therefore the only possible conclusion is $x = y$.

■

The statements

A : $a \leq b$ and $b \geq a$
and
B : $a - b = 0$

are equivalent. (Exercise)

There are at least two special ways to prove that a number is equal to zero. Both of them use the absolute value of a number.

Method 1. Since by definition of absolute value $|a| = 0$ if and only if $a = 0$, in order to prove that $a = 0$, we can prove that $|a| = 0$.

Method 2. We can use the following result:

Let a be a real number. $a = 0$ if and only if $|a| < \varepsilon$ for every real number $\varepsilon > 0$.

Proof: The statement "If $a = 0$, then $|a| < \varepsilon$ for every real number $\varepsilon > 0$" is clearly true. Indeed, if $a = 0$, $|a| = 0$.

Let's now consider its converse: "If $|a| < \varepsilon$ for every real number $\varepsilon > 0$, then $a = 0$."

(*Discussion:* By definition of "distance" on the real line, the distance between a and 0 is equal to $|a - 0| = |a|$. Since we can consider ε to be a real number as large or as small as we wish, the hypothesis suggests that, in particular, a must be as close to 0 as we wish. This information does not seem to offer any practical starting point. Let's try to prove the statement by contraposition.)

Thus we will assume that $a \neq 0$, and we will try to prove that there is at least one real number $\varepsilon > 0$ such that $|a|$ is not smaller than ε (or $|a| \geq \varepsilon$).

Since $a \neq 0$, $|a| > 0$. So $|a| \neq 0$.

Can we find a positive number ε, such that $|a| \geq \varepsilon$? If we could, this would imply that $0 < \varepsilon \leq |a|$.

Consider

$$\varepsilon = \frac{|a| + 0}{2} = \frac{|a|}{2}$$

Since $0 < \varepsilon < |a|$, we have found an $\varepsilon > 0$ such that $|a| \geq \varepsilon$.

This conclusion contradicts the original hypothesis "$|a| < \varepsilon$ for every real number $\varepsilon > 0$." So it is not possible that $a \neq 0$.

EXERCISES

Prove the following statements.

1. Let x and y be two real numbers.

$$(x-y)^5 + (x-y)^3 = 0 \text{ if and only if } x = y$$

2. Let x and y be two real numbers. The two sequences $\{x^n\}$ and $\{y^n\}$ (where n is a counting number, $n > 1$) are equal if and only if $x = y$.
   (A sequence is a collection of numbers ordered using the counting numbers.)
3. Let a, b, and c be three counting numbers. If a divides b, b divides c, and c divides a, then $a = b = c$.
4. Let a, b, and c be three counting numbers. Then

$$\text{GCD}(ac, bc) = c\,\text{GCD}(a,b).$$

5. Let a and b be two relatively prime integers. If there exists an m such that $(a/b)^m$ is an integer, then $b = 1$.

COMPOSITE STATEMENTS

The hypothesis and/or the conclusion of a proposition may be composite statements. Because of the more complicated structure of the statement, we have to pay more attention when reading it, but once the statement is clear the proof does not require any particular "trick."

Multiple Hypothesis

Multiple hypothesis statements are statements in which the hypothesis is a composite statement, such as "If A and B, then C."

The proof of this kind of statement is like that of "If A, then B." The only difference is that the composite hypothesis, A and B, probably contains more information than the proof in which the hypothesis is a simple statement.

Example 1: If b is a multiple of 2 and a multiple of 5, then b is a multiple of 10.

Proof:
A: b is a multiple of 2.
B: b is a multiple of 5.
(*Implicit hypothesis:* All the properties of integer numbers can be used)
C: b is a multiple of 10.

By A, b is divisible by 2. So $b = 2n$ for some integer n. The other hypothesis, B, says that b is divisible by 5 as well. Therefore $2n$ is divisible by 5.

By the laws of algebra a product is divisible by a prime number if at least one of its factors is divisible by that number. Since 2 is not divisible by 5, n must be divisible by 5. So $n = 5k$. Therefore

$$b = 2n = 2(5k) = 10k$$

This proves that b is divisible by 10.

■

The proof of a statement such as "If A and B, then C" can be done by contradiction, if needed. In this case we assume that C is false. We will get a contradiction if we can prove that the conclusion we obtain contradicts either A *or* B. We do not need to get a conclusion that contradicts both A *and* B; one contradiction is all we need.

Multiple Conclusion

The following are the most common kinds of multiple conclusion statements:

1. If A, then B and C are the only possible outcomes.
2. If A, then both B and C must occur.
3. If A, then B or C.
4. If A, then B or C cannot occur at the same time.

Let's look at these statements.

1. If A, then B and C are the only possible outcomes.

The statement says that under the given hypothesis there are only two possible outcomes.

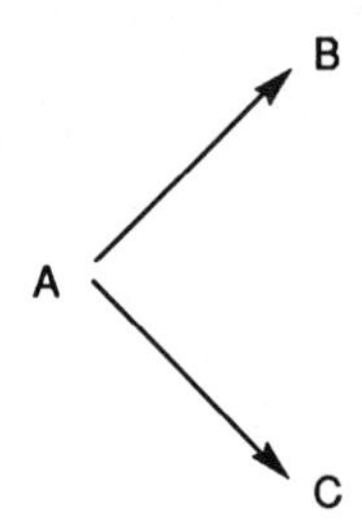

The proof uses "partial contradiction." Let's assume that A is true and outcome B is not possible (B is false).

If it is true that B and C were the only two possible outcomes, this will force C to be true.

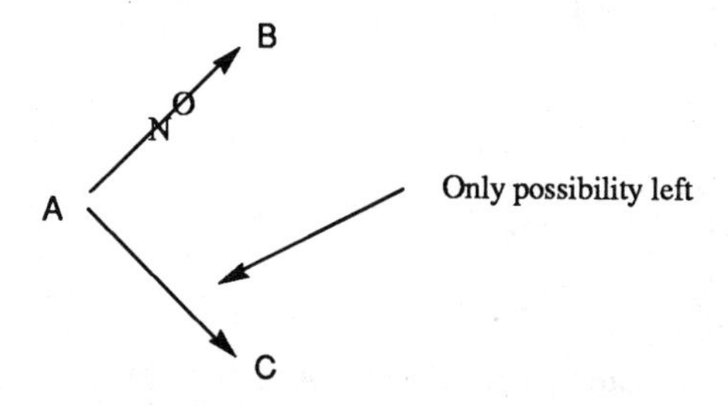

To describe this situation we changed the original statement into:

If A and (not B), then C.

(Or: If A and [not C], then B. In this case we chose to decide that outcome C could not be possible.)

After rewriting the statement, we can prove it by using one of the techniques seen before.

Example 2: If n is a positive integer, then either n is a multiple of 2 or n divided by 2 has remainder of 1.

Proof:

A: n is a positive integer (so we can use all the properties of integer numbers).

B: n is a multiple of 2.

C: n divided by 2 has remainder of 1.

Let's assume that n is an integer and that the remainder of the division between n and 2 is not 1 (composite hypothesis: "A and not C").

We have to show that n is a multiple of 2; that is, $n = 2q$ for some integer q. (In this way we prove that B must be true.)

The division algorithm establishes that the only possible remainder when dividing by 2 is 0 or 1. By hypothesis the remainder is not 1. So it must be 0.

Therefore $$n = 2q + 0 = 2q.$$

This means that n is a multiple of 2.

■

If we want to prove by contradiction a multiple conclusion statement like the one we are considering, we have to assume that the conclusion is false. Since the conclusion is the list of all the possible outcomes, we have to assume that *all* the possible outcomes are false. Therefore

If A, then either B or C

generates the following assumption as a starting point for a proof by contradiction:

"Assume that both B and C are false"

This statement should now enable us to get a conclusion that contradicts A.

2. The proof of the statement "If A, then both B and C" has two parts:

(i) If A, then B
and
(ii) If A, then C

Indeed, we have to show that each one of the possible conclusions is true, because we want both of them to hold.

If we have already proved that one of the two (or more) implications is true, we can use it to prove the remaining ones (if needed).

Example 3: The lines $y = 2x + 1$ and $y = -3x + 2$ are not perpendicular, and they intersect in exactly one point.

Proof:
A: The two lines have equations $y = 2x + 1$ and $y = -3x + 2$

We have to show that two conclusions hold true:

1. B: The lines are not perpendicular.
2. C: They intersect in exactly one point.

1. Two lines are perpendicular if and only if their slopes, m and m_1, satisfy the equation $m = -1/m_1$.
The first line has slope 2, the second has slope –3. Clearly

$$-3 \neq \frac{1}{-2}$$

So the lines are not perpendicular.

2. This second part is an existence and uniqueness statement: There is one and only one point belonging to both lines. We can prove this part in two ways:
a. The given lines are distinct and nonparallel (since they have different slopes); therefore they have only one point in common.
b. We can find the coordinates of the point by solving the system

$$y = 2x + 1$$

$$y = -3x + 2$$

By substitution we have

$$2x + 1 = -3x + 2$$

$$5x = 1$$

$$x = 1/5$$

The corresponding value of the y variable is

$$y = 2(1/5) + 1 = 7/5$$

Therefore the lines have the point with coordinates (1/5, 7/5) in common.
This point is unique because its coordinates represent the only solution of the system given by the equations of the two lines.

■

3. If A, then B or C.

In this case we need to show that given A, either B is true or C is true (not necessarily both). This means that we need to show that at least one of the possible conclusions is true.

Example 4: Let a be an even number, with $|a| > 16$. Then either $a \geq 18$ or $a \leq -18$.

Proof:
A: The number a is even and $|a| > 16$.
(*Implicit hypothesis:* a is an integer.)
B: $a \geq 18$
C: $a \leq -18$

Since $|a| > 16$, then we have two possible cases:

1. $a > 16$
2. $a < -16$

Let's assume that $a > 16$. The number a is even. Therefore it cannot be 17. Thus a must be at least 18; that is, $a \geq 18$. So in this case B is true.

Let's consider the other case, namely $a < -16$. Since a is even, it cannot be -17. Therefore $a \leq -18$. So in this case C is true.

■

4. If A, then B and C cannot occur at the same time.

Since the word *not* appears in the statement, we might get a hint at using the proof by contradiction.

What happens if we assume that B and C are true at the same time? If the statement we want to prove is true, we should obtain a conclusion that contradicts A.

Example 5: Let a be a nonzero positive integer. Then a cannot be even and odd at the same time.

Proof:
A: a is a nonzero positive integer number.
B: a is even.
C: a is odd.

Let's assume that a is even and odd at the same time.

Since it is even, a is a multiple of 2; that is, $a = 2p$ for some positive integer p. Since a is odd, it has remainder 1 when it is divided by 2.

Therefore $a = 2n + 1$ for some positive integer number n.
If we compare these two ways of writing a, we obtain

$$2p = a = 2n + 1$$

or

$$2p = 2n + 1$$

We can use the properties of real numbers (since integer numbers are real numbers), and subtract $2n$ from both sides of the last equality.

$$\begin{array}{r} 2p = 2n + 1 \\ \underline{-2n \quad -2n} \\ 2(p - n) = 1 \end{array}$$

The equality we have just obtained states that 1 is a multiple of 2, because the number $p - n$ is an integer (it is the difference between two integer numbers). From the properties of real numbers we know that this is not true. Therefore the assumption that a could be even and odd at the same time is wrong.

■

EXERCISES

Prove the following statements.

1. If $x^2 = y^2$ and $x \geq 0$, $y \geq 0$, then $x = y$.
2. If a function f is even and odd at the same time, then $f(x) = 0$ for all x in the domain of the function.
   (A function is called even if $f(x) = f(-x)$ for all x in its domain. It is called odd if $f(x) = -f(-x)$ for all x in its domain)
3. If n is a positive multiple of 3, then either n is odd or it is a multiple of 6.
4. If x and y are two real numbers such that $x^4 = y^4$, then either $x = y$ or $x = -y$.
5. Let A and B be two subsets of the same set U. Define

$$A - B = \{a \in A \mid a \notin B\}$$

If B is nonempty (that is, B has at least one element), $A \neq B$, and $A - B$ is empty, then either A is empty or $A \subset B$.

6. Fill in all the details and outline the following proof of the Rational Zero Theorem:

Let z be a rational zero of the polynomial

$$P(x) = a_n x^n + a_{n-1} x^{n-1} + \ldots + a_0$$

Suppose that $a_n, a_{n-1}, \ldots, a_1, a_0$ are integers with $a_n \neq 0$ and $a_0 \neq 0$, with $n \geq 1$. Let z be written in its lowest terms, $z = p/q$, with $q \neq 0$.
Then q divides a_n and p divides a_0.

Proof: By hypothesis $P(z) = 0$. So

$$a_n \left(\frac{p}{q}\right)^n + a_{n-1} \left(\frac{p}{q}\right)^{n-1} + \ldots a_1 \left(\frac{p}{q}\right) + a_0 = 0$$

Therefore

$$a_n p^n + a_{n-1} p^{n-1} q + \ldots + a_1 p q^{n-1} + a_0 q^n = 0 \qquad (*)$$

Thus we can solve for $a_n p^n$ and we obtain

$$a_n p^n = -q\,(a_{n-1} p^{n-1} + \ldots + a_1 p q^{n-2} + a_0 q^{n-1})$$

or $a_n p^n = -qt$, where $t = a_{n-1} p^{n-1} + \ldots + a_1 p q^{n-2} + a_0 q^{n-1}$ is an integer.
This implies that q divides $a_n p^n$.
Since p and q have no common factors, q cannot divide p^n. Thus q divides a_n. Part of the conclusion is therefore proved.
We can now solve (*) for $a_0 q^n$. Therefore

$$a_0 q^n = -p\,(a_n p^{n-1} + a_{n-1} p^{n-2} q + \ldots + a q^{n-1})$$

or $a_0 q^n = -ps$, where $s = a_n p^{n-1} + a_{n-1} p^{n-2} q + \ldots + a q^{n-1}$ is an integer.
Thus p divides $a_0 q^n$.
Since p and q have no common factors, p must divide a_0.

EXERCISES

Discuss the truth of the following statements. Prove the ones that are true; find a counterexample for each one that is false.

The exercises with the symbol (*) require knowledge of calculus and/or linear algebra.

1. Prove that if $P(x_1,y_1)$ and $Q(x_2,y_2)$ are two distinct points in the plane, then the distance between the two of them, defined as

$$d(P,Q) = \sqrt{(x_2 - x_1)^2 - (y_2 - y_1)^2}$$

is a positive number.

2. Let a be a real number. Then the opposite of a is unique.
3. Let n be any positive integer number. Then $\ln n < n$.
Prove this result in all of the following ways:
   a. By induction
   b. By graphing the two functions $f(x) = \ln x$ and $g(x) = x$ and comparing them. We can consider $x \geq 1$, since we are interested only in positive integers.
   c. By studying the function $h(x) = (\ln x)/x$ for $x \geq 1$. (*)
   (Show that the function is bounded between 0 and 1.)
   d. By studying the function $g(x) = \ln x - x$. (*)
   (Show that $g(x) < 0$ for all $x \geq 1$.)
4. a. Are these two sets equal?

   A = {all the integers that are multiples of 15}
   B = {all the integers that are multiples of 3 and of 5}

   b. Are these two sets equal?

   A = {all the integer numbers that are multiples of 15}
   B = {all the integer numbers that are either multiples of 3 or multiples of 5}

5. Let a and b be two real numbers with $a \neq 0$. Prove that the solution to the equation $ax = b$ exists and that it is unique.
6. The counting number n is odd if and only if n^3 is odd.
7. The following statements are equivalent:

   1) $a \leq b$ and $a \geq b$
   2) $a - b = 0$

8. Every nonzero real number has a unique reciprocal.
9. Let p, q, and n be three positive integers. If p and q have no common

factors, then q does not divide p^n.

10. Prove that for every integer $n > 0$

$$\frac{1}{1}\frac{1}{2}+\frac{1}{2}\frac{1}{3}+\frac{1}{3}\frac{1}{4}+\dots+\frac{1}{n}\frac{1}{n+1}=\frac{n}{n+1}$$

11. $\sqrt{2}$ is an irrational number.

12. Prove algebraically that two distinct lines have at most one point in common.

13. All negative numbers have negative reciprocals. (See Exercise 8.)

14. The remainder of the division of a polynomial $P(x)$ by $x - a$ is the number $P(a)$.

15. Let $P(x)$ be a polynomial of degree larger than or equal to one. The following statements are equivalent:

1. $x = a$ is a root of $P(x)$.
2. $P(x)$ can be exactly divided by $x - a$.
3. $x - a$ is a factor of $P(x)$.

16. Let f be a differentiable function at the point $x = a$. Then f is continuous at that point. (*)

17. All prime numbers larger than 2 are odd. (Use proof by contraposition.)

18. Let A be a 2x2 matrix. The following statements are equivalent:: (*)

1. A has an inverse.
2. Its determinant is nonzero.
3. The system $A\,[x\ y] = [0\ 0]$ has only the trivial solution $x = 0$, $y = 0$.

19. For all positive integer numbers k

$$1^3 + 2^3 + 3^3 + \dots + k^3 = k^2(k+1)^2/4$$

20. Let a and b be two real numbers. If $ab = (a+b)^2/4$, then $a = b$.

21. Let a and b be two real numbers. If $ab = (a+b)^2/2$, then $a = b = 0$.

22. For $k \geq 2$,

$$\frac{1}{k+1}+\frac{1}{k+2}+\dots+\frac{1}{2k}>\frac{1}{2}$$

23. Let a, b, and c be three integers. If a is a multiple of b and b is a multiple of c, then a is a multiple of c.

24. Let p be a nonzero real number. p is rational if and only if its reciprocal is a rational number.

25. Let a, b, and c be any three consecutive integers. Then 3 divides $a + b + c$.
26. Let k be a whole number. Then $k^3 - k$ is divisible by 3. How does this exercise relate to the previous exercise?
27. Let $\{a_n\}$ $n = 1, 2, \ldots$ be a sequence of real numbers converging to the number L; that is, $\lim a_n = L$. If $a_n > 0$ for all n, then $L \geq 0$. (*)
(We say that $\lim a_n = L$ if for every $\varepsilon > 0$ there exists an N such that $|a_n - L| < \varepsilon$ for all $n > N$.)
Can we say that if $a_n > 0$ for all n, then $L > 0$?
28. If $ad - bc$ is nonzero, then the system

$$\begin{aligned} ax + by &= e \\ cx + dy &= f \end{aligned}$$

has a unique solution.
29. Let $k \geq 6$. Then $2^k > (k + 1)^2$.
30. There exists a number k such that $2^k > (k + 1)^2$.
31. If t is a rational number and q is a rational number, then $t + q$ is an irrational number.
32. There are three consecutive numbers a, b, and c such that $a + b + c$ is divisible by 3.
33. If n is a positive multiple of 5, then n^2 is a multiple of 125.
34. For every integer n, the number $n^2 + n$ is always even.
35. Let $k \geq 6$. Then $k! > k^3$.

Solutions to the Exercises at the End of Each Section

WARNING

The following solutions to the exercises listed in this book should be used as a guide. Indeed, learning to construct proofs is like learning to play tennis. It is useful to have someone teaching us the basics and to look at someone playing, but we need to get into the court and play if we really want to learn.

Therefore we suggest that you, the reader, set aside a minimum time limit for yourself to construct a proof without looking at the solution (as a starting point, you could give yourself at least one hour, and then adjust this limit to fit your ability).

If you do not succeed, read only the first few lines of the proof presented here, and then try again on your own. If you cannot complete the proof, read a few more lines and try once more. If you need to read the whole proof, make sure that you understand it, and after a few days try the exercise again on your own.

The discussion part of a proof is enclosed in parentheses.

SOME BASIC TECHNIQUES USED IN PROVING A THEOREM OF THE FORM "A IMPLIES B"

1. There exists at least one real number for which the function f is not defined.

2. Given the two numbers x and y, there is no rational number z such that

$x + z = y$. (Or: Given the two numbers x and y, $x + z \neq y$ for all rational numbers z.)

3. There exist at least two distinct real numbers x and y for which $f(x) = f(y)$.

4. The equation $P(x) = 0$ has at least two solutions.

5. There is at least one nonzero number that has zero opposite.

6. Either: (i) There exists a number $n > 0$ for which there is not a number $M_n > 0$ such that $f(x) > n$ for all numbers x with $x > M_n$; or (ii) There exists a number $n > 0$ such that for every positive number M there is at least one x with $x > M$ and $f(x) \leq n$.

7. There exists at least one number satisfying the equation $P(x) = Q(x)$ such that $|x| \geq 5$.

8. Compare this statement to statement 4. In this case we do not know whether a solution exists at all. So the opposite of the statement should be: Either the equation $P(x) = 0$ has no solution, or it has two or more solutions.

9. The function f is not continuous at the point c if there exists an $\varepsilon > 0$ such that for all $\delta > 0$ there exists an x with $|x - c| < \delta$ and

$$|f(x) - f(c)| \geq \varepsilon$$

10. There exists at least one real number x_0 such that $f(x_0)$ is an irrational number.

1. Let x and y be any two real numbers such that $x \leq y$.
   Can we prove that $f \circ g(x) \leq f \circ g(y)$?
   Since g is nondecreasing, $s = g(x) \leq g(y) = t$.
   Since f is nondecreasing, $f(s) \leq f(t)$; that is, $f(g(x)) \leq f(g(y))$.
   So the statement is true.

2. (Since we cannot directly check all the products between rational and irrational numbers, we will use proof by contraposition.)
   Let's assume that xy is rational. Then $xy = a/b$ with a and b integers

and $b \neq 0$. Since x is rational, $x = c/d$ with c and d integers, with $d \neq 0$ and $c \neq 0$ (since x is a nonzero number). What can we find out about y?

Because of our assumptions, $\frac{c}{d}y = \frac{a}{b}$.

We can multiply both sides of the equality by d, since d is a nonzero number, and we obtain $cy = ad/b$. Since $c \neq 0$, we can divide both sides of the equality by c. Thus $y = ad/bc$.

Therefore y is rational. This is a contradiction because y is irrational by hypothesis. So xy must be irrational.

3. Since n has at least three digits, it looks like $n = rst \ldots cba$, where $r, s, t, \ldots, b$, and a represent its digits.

   Since we have some information about the last two digits of n, we will isolate them. Therefore we can write $n = (rst.... c)\,100 + ba$. By hypothesis ba is divisible by 4, so $ba = 4t$ for some integer number t. Thus $n = (rst... c)\,100 + 4t = 4[25(rst... c) + t\,]$.

   This proves that n is divisible by 4.

4. By hypothesis $(a + b)^2 = a^2 + b^2$. Since $(a + b)^2 = a^2 + 2ab + b^2$, we obtain $a^2 + 2ab + b^2 = a^2 + b^2$ for all real numbers b. This implies that $2ab = 0$ for all real numbers b.

   In particular the equality is true when $b \neq 0$. Thus we can divide $2ab = 0$ by $2b$ and we obtain $a = 0$.

5. (Since it is impossible to check directly all the prime numbers that can be written in the form $2^n - 1$ to see whether the exponent n is indeed a prime number, we will try to use proof by contraposition.)

   Let's assume that n is not a prime number. Then n is divisible by a number t different from n itself and from 1. So $t \geq 2$. Thus $n = tq$, where q is some positive integer, $q \neq 0$ (because $n \neq 0$), $q \neq 1$ (because $t \neq 1$), $q \neq n$ (why?).

   So we can write

$$2^n - 1 = 2^{tq} - 1 = (2^q)^t - 1^t$$

We can now use factorization techniques to obtain

$$2^n - 1 = (2^q - 1)\,[(2^q)^{t-1} + (2^q)^{t-2} + \ldots + 1]$$

This equality shows that $2^n - 1$ is not a prime number because it can be written as a product of two numbers, and neither of these two numbers is 1. (Why? Look at the information we have about t and q.)

Therefore n must be a prime number.

6. Let n be a four-digit palindrome number. In order to prove that n is divisible by 11 we can show that $n = 11t$ for some positive integer t.

   Since n is a four-digit palindrome number, we can write n as $xyyx$ with x and y integer numbers from 0 to 9 (x cannot equal 0, because it is the first digit of n). Since we have some information about the digits of n, we could try to separate them. Thus

$$n = xyyx = 1000x + 100y + 10y + x = 1001x + 110y = 11(91x + 10y)$$

   Since $t = 91x + 10y$ is an integer number, we proved that n is a multiple of 11.

7. We know that $x \neq c$. So the denominator of the fraction is a nonzero number, and the fraction is well defined.

   We want to prove that the number

$$\frac{f(c) - f(x)}{c - x}$$

   is nonnegative.

   (Because of the algebraic properties that determine the sign of a fraction, we need to prove that the numerator and the denominator are either both nonnegative or both nonpositive. Let's consider all possible cases.)

   Since $c - x \neq 0$, there are two possible cases:

   (i) $c - x > 0$
   (ii) $c - x < 0$.

   In the first case, namely $c - x > 0$, $c > x$. Since f is a nondecreasing function, $f(c) \geq f(x)$.

   So $f(c) - f(x) \geq 0$ and $c - x > 0$. This implies that the fraction

$$\frac{f(c) - f(x)}{c - x}$$

   is nonnegative. In the second case, namely $c - x < 0$, $c < x$. Since f is a nondecreasing function, $f(c) \leq f(x)$.

   So $f(c) - f(x) \leq 0$ and $c - x < 0$. This implies that the fraction

$$\frac{f(c) - f(x)}{c - x}$$

is nonnegative.

8. To prove that f is one-to-one, we have to prove that if $x_1 \neq x_2$, then $f(x_1) \neq f(x_2)$.
(i) *Direct method.* Let $x_1 \neq x_2$. Since $m \neq 0$, $mx_1 \neq mx_2$. Thus $mx_1 + b \neq mx_2 + b$. So $f(x_1) \neq f(x_2)$.
(ii) *Contraposition.* Let $x_1 \neq x_2$, and let's assume that $f(x_1) = f(x_2)$. This assumption allows us to set up an equality and use it as a starting point Then $mx_1 + b = mx_2 + b$. This equality implies that $mx_1 = mx_2$. Since $m \neq 0$, we can divide both sides of the equality by m, and we obtain $x_1 = x_2$. We obtained a contradiction, because $x_1 \neq x_2$.
Thus it is not possible to have $x_1 \neq x_2$ and $f(x_1) = f(x_2)$. So f is a one-to-one function.

9. We have to prove that for every real number y there is at least one real number x such that $f(g(x)) = y$.
By hypothesis f is onto. So there is at least one real number z such that $f(z) = x$. The function g is onto. So there is at least one real number t such that $g(t) = z$.
Therefore $f(g(t)) = f(z) = y$.

"IF AND ONLY IF" OR EQUIVALENCE THEOREMS

1. (a) If x is an integer not divisible by 2, then x is not divisible by 6. (b) If x is an integer divisible by 2, then x is divisible by 6. (c) If x is an integer not divisible by 6, then x is not divisible by 2.

2. (a) If the diagonals of a quadrilateral bisect, the quadrilateral is a parallelogram. (b) If the diagonals of a quadrilateral do not bisect, the quadrilateral is not a parallelogram. (c) If a quadrilateral is a parallelogram, then its diagonal bisect.

3. (a) If there is at least one i for which $a_i \neq b_i$, then there is at least one real number x for which $P(x)$ and $Q(x)$ are not equal. (b) If $a_i = b_i$ for all i, with $0 \leq i \leq n$, then $P(x)$ and $Q(x)$ are equal for all real numbers. c) If there exists at least one real number for which $P(x)$ and $Q(x)$ are not equal, then there is at least one i for which $a_i \neq b_i$.

1. *Part 1*

Assume that f is a nonincreasing function. We want to show that $(f(c) - f(x)) / (x - c) \leq 0$ for all c and x in the domain of f with $x \neq c$.

Since $x \neq c$, there are two possibilities: (a) $x < c$ or (b) $x > c$.

If $x < c$, it follows that $f(x) \geq f(c)$. These two inequalities can be rewritten as:

$$\text{if } x - c < 0, \text{ then } f(x) - f(c) \geq 0$$

Therefore $(f(c) - f(x)) / (x - c) \leq 0$.

If $x > c$, it follows that $f(x) \leq f(c)$. These two inequalities can be rewritten as:

$$\text{if } x - c > 0, \text{ then } f(x) - f(c) \leq 0$$

Therefore $(f(c) - f(x)) / (x - c) \leq 0$.

Part 2

Our hypothesis is $(f(c) - f(x)) / (x - c) \leq 0$ for all c and x in the domain of f with $x \neq c$. Does this inequality imply that f is a nonincreasing function?

A quotient between two real numbers is nonpositive when one of the two numbers is negative and the other is positive (or the dividend is zero).

Suppose that the denominator is positive. Then the numerator must be either negative or zero. This means that if $x - c > 0$ we must have $f(x) - f(c) \leq 0$.

Therefore if $x > c$ it follows that $f(x) \leq f(c)$.

Suppose that the denominator is negative. Then the numerator is either positive or zero. This means that if $x - c < 0$, then $f(x) - f(c) \geq 0$; or, if $x < c$, then $f(x) \geq f(c)$.

Therefore f is a nonincreasing function.

2. We will show that (1) implies (2), (2) is equivalent to (3), and (3) implies (1). (This is only one of the possible ways to prove the equivalence of these statements.)

(1) implies (2):

We have to prove that $a^s < a^r$ for all real numbers $a > 1$; or, equivalently, that $a^s - a^r < 0$ for all real numbers $a > 1$.

Since $r > s$, we can write $a^s - a^r = a^s(1 - a^r - s) = a^s(1 - a^t)$ where t is the positive number $r - s$.

Let's consider the two factors of this product: a^s and $1 - a^t$. The first one, a^s, is positive, since a is > 1. The second one is negative because $a > 1$, so $a^t > 1$. Therefore $a^s(1 - a^t) < 0$. This proves the conclusion.

(2) implies (3):

We have a number $a < 1$. The hypothesis is about numbers > 1. Is there a relationship between these two kinds of numbers? Yes. Any number smaller than 1 is the inverse of a number larger than 1:

$$a = (a^{-1})^{-1} = b^{-1} \text{ with } b > 1$$

Then by hypothesis $b^s < b^r$. Therefore $(a^{-1})^s < (a^{-1})^r$. This implies that $a^s > a^r$.

(3) implies (2):

Any number larger than 1 is the inverse of a number smaller than 1:

$$a = (a^{-1})^{-1} = b^{-1} \text{ with } b < 1$$

Then by hypothesis $b^r < b^s$. Therefore $(a^{-1})^r < (a^{-1})^s$.

This implies that $a^s < a^r$.

(3) implies (1):

By hypothesis $a^r < a^s$ for all real numbers $a < 1$, or $a^r - a^s < 0$ for all real numbers $a < 1$.

We can rewrite this product as $a^r(1 - a^{s-r})$. The product is negative and one of its factors, a^r, is positive. Therefore the other factor, $1 - a^{s-r}$, must be negative. So $1 < a^{s-r}$.

Since $a < 1$, this will happen if $s - r < 0$. Therefore we get $s < r$.

3. (i) implies (ii): Already proved.

(ii) implies (i): From the inequality $(a + b)/2 > a$ we obtain $a + b > 2a$.

This implies $b > a$.

(i) implies (iii): $b > a$. Therefore $a + b > a + a = 2a$.

This inequality implies $(a + b)/2 > a$.

(iii) implies (i): Already proved.

4. (i) implies (ii): The numbers x and y are both negative. Thus, by the definition of absolute value $|x| = -x$, and $|y| = -y$. The inequality $x < y$ implies $-x > -y$.

So $|x| > |y|$.

(ii) implies (i): Again, since x and y are negative numbers, $|x| = -x$, and $|y| = -y$. Since $|x| > |y|$, it follows $-x > -y$. This last inequality implies $x < y$.

(iii) implies (i): By hypothesis, $x^2 > y^2$. So $x^2 - y^2 > 0$, or

$$(x - y)(x + y) > 0$$

The number $x + y$ is negative because x and y are both negative numbers. The product $(x - y)(x + y)$ can be positive only if the number $x - y$ is negative.

So $x - y < 0$, or $x < y$.

(i) implies (iii): Since $x < y$, $x - y < 0$.

We must gather information about the number $x^2 - y^2$.

Using factorization techniques we can write

$$x^2 - y^2 = (x - y)(x + y)$$

The number $x - y < 0$, by hypothesis. The number $x + y < 0$, because it is the sum of two negative numbers. So their product is positive.

Therefore $x^2 - y^2 > 0$, or $x^2 > y^2$.

The theorem is now completely proved.

5. Assume that (x_0,y_0) is a solution of S_1. Is it a solution of S_2?

By definition of solution, (x_0,y_0) satisfies both equations in system S_1. So (x_0,y_0) satisfies the first equation of S_2 as well. Thus we must prove that it satisfies the second equation of S_2.

$$(a_1 + ba_2)x_0 + (b_1 + bb_2)y_0 = (a_1x_0 + b_1y_0) + b(a_2x_0 + b_2y_0)$$

Since (x_0,y_0) is a solution of S_1, $a_1x_0 + b_1y_0 = c_1$, $a_2x_0 + b_2y_0 = c_2$. Thus $(a_1 + ba_2)\, x_0 + (b_1 + bb_2)\, y_0 = c_1 + bc_2$. So (x_0,y_0) is a solution of S_2.

Assume that (x_0,y_0) is a solution of S_2. Is it a solution of S_1? By definition of solution, (x_0,y_0) satisfies both equations in system S_2. So (x_0,y_0) satisfies the first equation of S_1 as well. Thus we must prove that it satisfies the second equation of S_1.

By hypothesis $(a_1 + ba_2)\, x_0 + (b_1 + bb_2)\, y_0 = c_1 + bc_2$. We can rewrite the left-hand side of the second equation of S_2 as:

$$(a_1 + ba_2)x_0 + (b_1 + bb_2)y_0 = (a_1x_0 + b_1y_0) + b(a_2x_0 + b_2y_0)$$

Since (x_0,y_0) is a solution of S_2, the left-hand side of the equation is equal to $c_2 + bc_2$, and the first expression in parentheses in the right-hand side is equal to c_1. So we obtain

$$c_1 + bc_2 = c_1 + b(a_2x_0 + b_2y_0)$$

This equality implies $a_2x_0 + b_2y_0 = c_2$ because b is a nonzero number. This proves that (x_0,y_0) is a solution of S_1.

USE OF COUNTEREXAMPLES

1. (Let's consider the statement. It seems to suggest that the "growth" of f should be cancelled by the "decrease" of g. But f and g could increase and decrease at different rates. The statement does not seem to be true.)

   Let's look for a counterexample. Let's try to use easy functions, such as linear functions.

   Take $f(x) = x + 1$ and $g(x) = -3x$. Clearly, f is increasing and g is decreasing. Their sum is $h(x) = -2x + 1$, which is decreasing.

2. There might be an angle in the first quadrant for which $2 \sin t = \sin 2t$. But the equality is not true for all the angles in the first quadrant. Take $t = \pi/4$. Therefore

   $$2\sin t = 2(1/\sqrt{2}) = \sqrt{2}$$

   The other side of the equality gives $\sin\pi/2 = 1$. Thus the statement is false.

3. (It might seem plausible that $y = P(x)$ is always negative, because its leading coefficient is -1. But when the value of the variable x is "small," the value of the monomial $-x^2$ is smaller than the value of the monomial $2x$. This remark seems to suggest that for "small," positive values of x the variable $y = P(x)$ might be positive [or at least nonnegative].)

nonnegative. $P(1) = -(1)^2 + 2(1) - (3/4) = 1/4$.

Another way to prove that the statement is false is to construct the graph of the polynomial $P(x)$ and to observe that the graph is not completely located below the x-axis.

4. The statement seems to be true. But if $x = 1$, then $y = 1$.

   So we found a counterexample. The statement becomes true if we change either the hypothesis to "The reciprocal of a number $x > 1$" or the conclusion to "$0 < y \leq 1$."

5. If $n = 1$, then $3^1 + 2 = 5$, which is a prime number.
   If $n = 2$, then $3^2 + 2 = 11$, which is a prime number.
   If $n = 3$, then $3^3 + 2 = 29$, which is a prime number.
   If $n = 4$, then $3^4 + 2 = 83$, which is a prime number.
   If $n = 5$, then $3^5 + 2 = 245$, which is not a prime number. So the statement is false.

6. (The functions $f \circ g$ and $f \circ h$ are equal if and only if $f \circ g(x) = f \circ h(x)$ for all values of the variable x.

   By definition of composition of functions, this equality can be rewritten as $f(g(x)) = f(h(x))$. Can we conclude that $g(x) = h(x)$? Or could we find a function f such that $f(g(x)) = f(h(x))$ even if $g(x) \neq h(x)$?

   Let's look for some simple functions that might give a counterexample. [Keep in mind that counterexamples need not be "difficult"].)

   Let's try to use $g(x) = x$ and $h(x) = -x$. Is it possible to choose a function f such that $f(g(x)) = f(h(x))$?

   Given our choices for the functions g and h, this equality becomes $f(x) = f(-x)$. So we need to find a function f that assigns the same values to the number x and to its opposite, $-x$. What about $f(x) = x^2$?

   Let's check whether we found a counterexample.

$$f \circ g(x) = f(g(x)) = f(x) = x^2$$

$$f \circ h(x) = f(h(x)) = f(-x) = (-x)^2 = x^2$$

   So the equality $f(g(x)) = f(h(x))$ holds, but the functions g and h are not equal.

   Using the same choices for g and h, we could have used $f(x) = \cos x$, or $f(x) = x^4$, or $f(x) = x^6$, and so on.

7. Let's try to prove this statement. Let n be the smallest of the five consecutive integers we are going to add. Then the other four numbers can

be written as $n + 1, n + 2, n + 3$, and $n + 4$.

The sum of these five numbers is

$$S = n + (n + 1) + (n + 2) + (n + 3) + (n + 4) = 5n + 10$$

S is clearly divisible by 5, since $S = 5(n + 2)$.

The statement is therefore true.

8. Let's try to construct a proof of the statement.

The inequality $f(x) \leq g(x)$ is equivalent to the inequality $f(x) - g(x) \leq 0$. Therefore we can concentrate on proving that $f(x) - g(x) \leq 0$ for all real numbers $x \geq 0$.

$$f(x) - g(x) = x^2 - x^4 = x^2(1 - x^2) = x^2(1 - x)(1 + x)$$

Is this product smaller than or equal to 0 for all real numbers $x \geq 0$?

The product is equal to 0 for $x = 0$ and $x = 1$. (We are not considering $x = -1$ because x must be a nonnegative number)

Let's check what happens to the product $x^2(1 - x)(1 + x)$ if x is neither 0 nor 1. The number x^2 is always positive.

We know that $x \geq 0$, so $1 + x \geq 1$. Thus the product is smaller than or equal to 0 if and only if $1 - x \leq 0$. This implies $1 \leq x$.

Therefore $x^2(1 - x)(1 + x) \leq 0$ only when $1 \leq x$. So $f(x) - g(x) \leq 0$ only if $1 \leq x$, and not for all $x \geq 0$.

Thus the statement is false. Can we find a counterexample?

Take $x = 0.2$. Then $x^2 = .04$ and $x^4 = 0.0016$. Therefore in this case $x^2 > x^4$, and the statement is false.

9. Let n be the smallest of the four counting numbers we are considering. Then the other three numbers are $n + 1, n + 2$, and $n + 3$.

Let's add the four numbers. We obtain

$$S = n + (n + 1) + (n + 2) + (n + 3) = 4n + 6$$

The number S is not *always* divisible by 4. Indeed if $n = 1, S = 10$.

So we found a counterexample. The sum of the four consecutive integers 1, 2, 3, and 4 is not divisible by 4.

The given statement is false.

10. (This statement seems to be similar to the statement "The sum of two odd numbers is an even number." But similarity is never a proof, and statements that sound similar can have very different meanings. So we must try to construct a proof.)

To prove that the function $f + g$ is even we need to prove that

$$(f + g)(x) = (f + g)(-x) \text{ for all real numbers}$$

We know that $(f + g)(x) = f(x) + g(x)$ by definition of $f + g$. Again by definition, $(f + g)(-x) = f(-x) + g(-x)$. We will now use the fact that f and g are odd functions.

Therefore $f(-x) = -f(x)$ and $g(-x) = -g(x)$. So we obtain

$$(f + g)(-x) = f(-x) + g(-x) = [-f(x)] + [-g(x)]$$

$$= -[f(x) + g(x)] = -(f + g)(x)$$

Thus $(f + g)(-x) = -(f + g)(x)$, and $-(f + g)(x)$ in general is not equal to $(f + g)(x)$. So the equality $(f + g)(x) = (f + g)(-x)$ does not seem to be true for all real numbers.

Can we find a counterexample? The functions $f(x) = x$ and $g(x) = 2x$ are two odd functions (check this claim). Their sum is the function $(f + g)(x) = 3x$, which is not even. (Moreover, $f + g$ is odd).

11. The function f/g is not going to be defined for all real numbers, in general. Indeed, it is not defined for all values of x that are zeros for the function g.

So in general this statement is false. The fact that f/g is either even or odd is not relevant.

As a counterexample, consider $f(x) = x$ and $g(x) = x^3 - x^5$.

Try to prove the following statement: Let f and g be two odd functions, defined for all real numbers. Their quotient f/g defined as

$$\frac{f}{g}(x) = \frac{f(x)}{g(x)}$$

is an even function defined for all real numbers for which $g(x) \neq 0$.

MATHEMATICAL INDUCTION

1. a. Is the statement true for $k = 1$?

$$1 = 2^1 - 1$$

b. Let's assume that the equality is true for $k = n$. So

$$1 + 2 + 2^2 + 2^3 + \ldots + 2^{n-1} = 2^n - 1$$

c. Let's check whether the equality is true for $n + 1$:

$$1 + 2 + 2^2 + 2^3 + \ldots + 2^{n-1} + 2^{(n+1)-1} = 2^{n+1} - 1$$

or

$$1 + 2 + 2^2 + 2^3 + \ldots + 2^{n-1} + 2^n = 2^{n+1} - 1$$

$$1 + 2 + 2^2 + 2^3 + \ldots + 2^{n-1} + 2^n$$

$$= (1 + 2 + 2^2 + 2^3 + \ldots + 2^{n-1}) + 2^n \qquad \text{inductive hypothesis}$$

$$= (2^n - 1) + 2^n = 2(2^n) - 1 = 2^{n+1} - 1$$

So the statement is true.

2. Let's prove this statement by induction.

a. Let's assume that the statement is true for $k = 1$. Indeed, when $k = 1$, $9^k - 1 = 8$. And 8 is divisible by 8.

b. Let's assume it is true for $k = n$. So $9^n - 1 = 8q$ for some rational number q.

c. Prove the statement for $n + 1$.

$$9^{n+1} - 1 = 9^{n+1} - 1^{n+1} \qquad \text{use factorization techniques}$$

$$= (9 - 1)(9^n + 9^{n-1} + \ldots + 1)$$

$$= 8(9^n + 9^{n-1} + \ldots + 1) = 8s$$

with s an integer number.

Look carefully at the proof we just completed. Is there anything peculiar about it?

We did not use the inductive hypothesis at all. We used mathematical induction in a case in which it is not necessary.

Let's try to prove the given statement in a different way. The basic tool we used in the previous attempt is the factorization technique. The difference $9^k - 1$ can always be factored when $k > 1$.

Therefore we can separate the case $k = 1$ from the case $k > 1$.

If $k = 1$, as we already observed, $9^k - 1 = 8$.

If $k > 1$,

$$9^k - 1 = 9^k - 1^k \qquad \text{use factorization techniques}$$

$$= (9 - 1)(9^{k-1} + 9^{k-2} + \ldots + 1)$$

$$= 8(9^{k-1} + 9^{k-2} + \ldots + 1) = 8s$$

where s is an integer number. So 9^{k-1} is always divisible by 8.

This exercise shows you that *you* have to check which technique fits the statement you are dealing with, and there is nothing wrong with having to write a proof more than once.

3. (i) Let's check whether the statement is true for $k = 1$.

When $k = 1$, $2k = 2$. Therefore we have only one number in the left-hand side of the equation. We obtain

$$2 = 1^2 + 1$$

(ii) Assume that the statement is true for $k = n$; that is,

$$2 + 4 + 6 + \ldots + 2n = n^2 + n$$

(iii) Prove that the equality is true for $k = n + 1$.

The last number in the left-hand side of the equation is $2(n + 1) = 2n + 2$. So we must add all the even numbers between 2 and $2n + 2$.

The largest even number smaller than $2n + 2$ is $(2n + 2) - 2 = 2n$. (The difference between two consecutive odd numbers is 2.) Thus the equation we are going to check becomes

$$1 + 3 + 5 + \ldots + 2n + (2n + 2) = (n + 1)^2 + (n + 1)$$

$$1 + 3 + 5 + \ldots + 2n + (2n + 2) \qquad \text{associative property}$$

$$= [1 + 3 + 5 + \ldots + 2n\,] + (2n + 2) \qquad \text{inductive hypothesis}$$

$= n^2 + n + (2n + 2) = n^2 + 2n + 1 + n + 1$

$= (n + 1)^2 + (n + 1)$

So the given equality holds for all $k \geq 1$.

4. a. We have to check whether the statement is true for $k = 3$.

$$(1 + a)^3 = 1 + 3a + 3a^2 + a^3 > 1 + 3a^2$$

b. Let's assume that the inequality is true for $k = n$.

c. Let's check whether $(1 + a)^{(n+1)} > 1 + (n + 1)a^2$.

$$(1 + a)^{(n+1)} = (1 + a)^n(1 + a) > (1 + na^2)(1 + a)$$

$$= 1 + na^2 + a + na^3 > 1 + a^2 + na^3$$

Since $a > 1$, $a^3 > a^2$. So

$$(1 + a)^{(n+1)} > 1 + na^2 + na^3 > 1 + na^2 + na^2$$

$$> 1 + na^2 + a^2 = 1 + (n + 1)\, a^2$$

5. a. We will check the equality for $k = 1$.
In this case the left-hand side is only 1/2. The right-hand side is $[1 - (1/2)^2] / [1 - (1/2)] - 1 = 1/2$. So the equality is true.

b. Let's assume the inequality works for a number $k = n$.

c. We will try to prove that

$$1/2 + \ldots + (1/2)^n + (1/2)^{n+1} = [1 - (1/2)^{n+2}] / [1 - (1/2)] - 1$$

Using the inductive hypothesis

$$1/2 + \ldots + (1/2)^n + (1/2)^{n+1} = \{1/2 + \ldots + (1/2)^n\} + (1/2)^{n+1}$$

$$= \{[1 - (1/2)^{n+1}]/[1 - (1/2)] - 1\} + (1/2)^{n+1}$$

$$= [1 - (1/2)^{n+2}]/[1 - (1/2)] - 1$$

So the statement is true.

6. a. Check the inequality for $k = 3$.
In this case $3^2 = 9$ and $5(3!) = 5(6) = 30$. So the statement is true.

b. Let's assume that the inequality works for an arbitrary number $k = n$.

c. Is $(n+1)^2 \le 5(n+1)!$?

$$5(n+1)! = 5(n+1)\,n! = 5nn! + 5n! \ge$$

$$\ge 5nn! + n^2 \ge 5n(3) + n^2 \ge$$

$$\ge n^2 + 2n + n \ge n^2 + 2n + 1 = (n+1)^2.$$

So the inequality is true for all $n \ge 3$.

6. a. The statement is true for $n = 1$.

b. Assume it is true for all the odd numbers from 1 to n, with n odd.

c. Is the statement true for the next odd number, namely $n + 2$?

$$(n+2)^2 - 1 = (n^2 - 1) + 4(n+1)$$

Since $n^2 - 1 = 4m$ for some positive integer m (inductive hypothesis), we have

$$(n+2)^2 - 1 = 4(m + n + 1)$$

Thus $(n+2)^2 - 1$ is divisible by 4.

EXISTENCE THEOREMS

1. (We are trying to find a function defined for all real numbers. Usually polynomials are a good guess. But in general the range of a polynomial is much larger than the interval [0,1]. We could try to construct a rational function, because it is possible to use the denominator to control the growth of the function. But rational functions might not be defined for all real numbers. Why don't we try with trigonometric functions?)

The functions $\sin x$ and $\cos x$ are bounded, because $-1 \leq \sin x \leq 1$ and $-1 \leq \cos x \leq 1$. But they have negative values as well. We could try to square them or to consider their absolute values, so that we obtain positive, bounded functions.

The functions $\sin^2 x$, $\cos^2 x$, $|\sin x|$, $|\cos x|$ are functions defined for all real numbers and whose range is the interval $[0,1]$. You can graph them to check the claim, if you wish.

(The functions $\sin^2 x$ and $\cos^2 x$ are differentiable, while $|\sin x|$ and $|\cos x|$ are not differentiable at all points.)

2. Just take $n = 2$.

3. We are looking for a number b such that $ab = n$, where n is an integer. We know that $a \neq 0$, since 0 is a rational number. Then we can solve the equality and use $b = na^{-1}$.

   So, for example, $b = a^{-1}$ itself satisfies the requirement.

4. A second-degree polynomial $P(x)$ can be written as $P(x) = ax^2 + bx + c$, with a, b, and c real numbers, and $a \neq 0$.

   We are looking for a second-degree polynomial satisfying the requirements $P(0) = -1$ and $P(-1) = 2$. Therefore we obtain

   (i) $P(0) = a(0)^2 + b(0) + c = c$. So $c = -1$.
   (ii) $P(-1) = a(-1)^2 + b(-1) + c = a - b + c = 2$.

   If we use $c = -1$ in the second equation, we get $a - b = 3$.

   We have one equation, and two variables. Thus one of them will be used as a parameter. As an example, we can write $a = b + 3$ and set $b = 0$. We obtain the polynomial

$$P(x) = 3x^2 - 1$$

   This is a second-degree polynomial that satisfies the requirements $P(0) = -1$ and $P(-1) = 2$.

5. Since b^a will be a negative number, b must be negative. Indeed, powers of positive numbers are always positive. Moreover, a cannot be an even number, because an even exponent generates a positive result.

   The numbers a and b can be either fractions or integer numbers, so do not concentrate only on integer numbers.

   Let's try $b = -27$ and $a = 1/3$. Then $a^b = (1/3)^{-27} = 3^{27}$, which is a positive integer number, and $b^a = (-27)^{1/3} = -3$, a negative integer.

6. We can consider two cases: either $a_n > 0$ or $a_n < 0$.

Let's assume that $a_n > 0$.

When the variable x has a very large positive value, the value of $P(x)$ will be positive, because the leading term, a_nx^n, will overpower all the other terms.

When the variable x is negative, and very large in absolute value, the value of $P(x)$ will be negative (again the leading term, a_nx^n, will overpower all the other terms).

Since polynomials are continuous functions, by the Intermediate Value Theorem there exists a value of x for which $P(x) = 0$.

We can similarly prove that the statement is true for $a_n < 0$.

UNIQUENESS THEOREMS

1. For the sake of completeness we must show that:
(a) the polynomial $p(x)$ has a zero and (b) the zero is unique.

a. *Existence*. Find the value(s) of the variable x for which $p(x) = 0$.

To do so we have to solve the equation $x - b = 0$. Using the properties of real numbers we can obtain $x = b$.

b. *Uniqueness*. We can prove this in at least three ways.

1. We can use the result stating that a polynomial of degree n has at most n solutions and conclude that a polynomial of degree 1 has at most one solution. Since we have found the solution, it must be the only one.

2. The solution is unique because of the process used to find it.

3. We could assume that the number t is another zero of the polynomial $p(x)$, and check whether $t = b$, the solution we found in part a.

Since t is a zero of $p(x)$, $p(t) = 0$. On the other hand, $p(b) = 0$. Thus $p(t) = p(b)$.

This means $t - b = b - b$.

Adding b to both sides of the equality we obtain $t = b$.

2. In this case we have to prove existence and uniqueness of the solution of the equation $\cos\theta = \theta$ in the interval $[0,\pi]$.

One way of proving the statement is to graph the functions $f(x) = \cos x$ and $g(x) = x$. If the two graphs have only one intersection point in the interval $[0,\pi]$, the proof is complete.

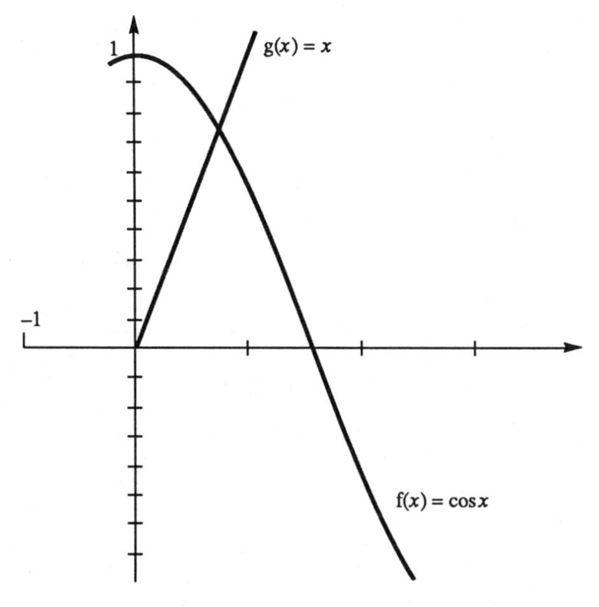

Another way to prove the statement is to graph the function $h(x) = \cos x/x$ and to show that there is only one value of x corresponding to $y = 1$. (Be careful: $h(x)$ is not defined at $x = 0$. Anyhow, when $x = 0$, $\cos x = 1$. So $\cos 0 \neq 0$.)

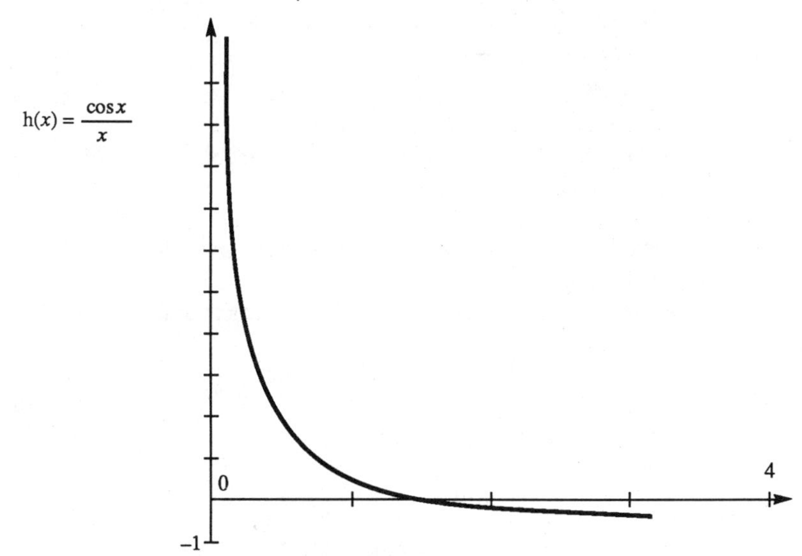

If you wish, you can try to solve algebraically the equation $\cos x = x$ and obtain a value t in the interval $[0,\pi]$.

3. We can start by finding a solution for the given equation and then proving that it is unique. Through algebraic manipulation we obtain that

$$x = \sqrt[3]{b}$$

is a solution. Moreover, since it is always possible to evaluate the cubic root of a real number, we can find x for any value of b.

Now, let y be another solution of the same equation. Then

$$x^3 - b = 0 = y^3 - b$$

This implies

$$x^3 - y^3 = 0$$

or

$$(x - y)(x^2 + xy + y^2) = 0$$

We now have two possibilities:

(1) $x - y = 0$ and
(2) $x^2 + xy + y^2 = 0$.

The first one implies that $x = y$ and the two solutions do indeed coincide. The second one is never true for real numbers x and y.

The expression $x^2 + xy + y^2$ is never equal to zero. Try to solve for x in terms of y, using the quadratic formula.

4. A second-degree polynomial is written as $P(x) = ax^2 + bx + c$, with a, b, and c real numbers, and $a \neq 0$.

We will use the conditions given in the statement to find a, b, and c.

$$P(0) = a(0)^2 + b(0) + c = c$$

$$P(1) = a(1)^2 + b(1) + c = a + b + c$$

$$P(-1) = a(-1)^2 + b(-1) + c = a - b + c$$

Thus

$$c = -1$$

$$a + b + c = 3$$

$$a - b + c = 2$$

We can solve this system and obtain

$$c = -1$$

$$a + b = 4$$

$$a - b = 3$$

So $a = 7/2$, $b = 1/2$, and $c = -1$. Thus the polynomial we are looking for is

$$P(x) = \frac{7}{2}x^2 + \frac{1}{2}x - 1$$

We have just proved that a polynomial satisfying the given requirements does exist.

Is this polynomial unique? The values of a, b, and c we obtained are the only solutions to the equations given by the conditions dictated in the statement. (For a precise proof of this statement, you might want to check a linear algebra textbook.) Therefore the polynomial we obtained is the only one satisfying the requirements.

5. Let's assume that there are at least two ways of writing n as the product of prime factors written in nondecreasing order. Therefore

$$p_1 p_2 \cdots p_k = n = q_1 q_2 \ldots q_s$$

Thus the prime factor p_1 divides $q_1 q_2 \ldots q_s$. This implies that p_1 divides at least one of the q_j. Let's assume that p_1 divides q_1 (we can reorder the q_1). Since q_1 is prime, this means $p_1 = q_1$.Therefore

$$p_2 \cdots p_k = q_2\, q_s$$

Similarly, p_2 divides $q_2 \ldots q_s$. So we can assume that p_2 divides q_2.

This again implies that $p_2 = q_2$. Thus

$$p_3 \cdots p_k = q_3 \cdots q_s$$

So if $k < s$ we obtain

$$1 = q_{k+1} \cdots q_s$$

But this is impossible since all the q_j are larger than 1.

If $k > s$ we obtain

$$1 = p_{s+1} \cdots p_k$$

Again this is impossible. Therefore $s = k$ and $p_1 = q_1, p_2 = q_2, \ldots, p_k = q_k$. So the factorization of n is unique for the prime numbers used. The order in which these factors are arranged is unique, since it is fixed. Therefore the factorization of n as described is unique.

EQUALITY OF SETS

1. First part: $(A \cup B) \cup (A \cup C) \subset A \cup (B \cup C)$

Let $x \in (A \cup B) \cup (A \cup C)$. Then either $x \in A \cup B$ or $x \in A \cup C$.

Thus either ($x \in A$ or $x \in B$) or ($x \in A$ or $x \in C$). If we eliminate the redundant part of this sentence, we can write it as either $x \in A$ or $x \in B$ or $x \in C$.

This implies that either $x \in A$ or $x \in B \cup C$. So $x \in A \cup (B \cup C)$.

Second part: $A \cup (B \cup C) \subset (A \cup B) \cup (A \cup C)$

Let $x \in A \cup (B \cup C)$. Then either $x \in A$ or $x \in B \cup C$. This implies either $x \in A$ or $x \in B$ or $x \in C$.

Therefore either ($x \in A$ or $x \in B$) or ($x \in A$ or $x \in C$). So we can conclude that $x \in (A \cup B) \cup (A \cup C)$.

2. First part: $A \subset B$

Let $x \in A$. Then x is a multiple of 2 and of 3. Therefore $x = 2n$.

Since x can be divided by 3 as well, while 2 is not, we can conclude that n is divisible by 3. So $x = 2(3m) = 6m$. Therefore x is divisible by 6.

Second part: $B \subset A$

Let $x \in B$. Then x is a multiple of 6. Thus we can write $x = 6t$ for some integer number t. Then x can be divided by 2 and 3, because 6 is divisible by 2 and 3.

So $x \in A$.

3. 1. $(A \cup B)' \subset A' \cap B'$

Let $x \in (A \cup B)'$. This implies that $x \notin A \cup B$. Therefore $x \notin A$ and $x \notin B$ (because if x belonged to either A or B, then it would belong to their union).

Thus $x \in A'$ and $x \in B'$. This implies that $x \in A' \cap B'$.

2. $A' \cap B' \subset (A \cup B)'$

Let $x \in A' \cap B'$. Then $x \in A'$ and $x \in B'$. Therefore $x \notin A$ and $x \notin B$. This implies that $x \notin A \cup B$. So we can conclude that $x \in (A \cup B)'$.

4. 1. $(A \cap B) \cap C \subset A \cap (B \cap C)$

Let $x \in (A \cap B) \cap C$. Thus $x \in (A \cap B)$ and $x \in C$.

This implies that $x \in A$ and $x \in B$, and $x \in C$. So $x \in A$, and $x \in (B \cap C)$. Therefore $x \in A \cap (B \cap C)$.

2. $A \cap (B \cap C) \subset (A \cap B) \cap C$

Let $x \in A \cap (B \cap C)$. Thus $x \in A$ and $x \in (B \cap C)$.

This implies that $x \in A$ and $x \in B$ and $x \in C$. So $x \in (A \cap B)$ and $x \in C$. Therefore $x \in (A \cap B) \cap C$.

5. The two sets are not equal. The number 144 is in A, since $144 = 16(9)$, and $144 = 36(4)$, but 144 is not in B.

6. 1. If you wish to do so, you can use a Venn diagram to get a better grasp of the set involved.

The equality does not seem to be true in general, since it contradicts one of the distributive laws. Let's look for a counterexample.

If $A = \{1\}$, $B = \{2\}$, and $C = \{2,3\}$, then

$$A \cup (B \cap C) = \{1,2\}$$

$$(A \cup B) \cap C = \{2\}$$

Therefore the two sets are not equal, in general.

2. This equality seems to be an extension of one of De Morgan's laws. Let's try to prove it.

The element x belongs to $(A \cap B \cap C)'$ if and only if $x \notin A \cap B \cap C$.

This happens if and only if either $x \notin A$, or $x \notin B$, or $x \notin A$. This is equivalent to saying that either $x \in A'$, or $x \in B'$, or $x \in C'$.

Thus $x \in A' \cup B' \cup C'$.

7. The element (x_0, y_0) belongs to A if and only if $y_0 = x_0^2 - 1$. This equality is equivalent to

$$y_0 = (x_0^2 - 1)\,\frac{x_0^2 + 1}{x_0^2 + 1}$$

Therefore $y_0 = x_0^2 - 1$ if and only if $y_0 = (x_0^4 - 1)/(x_0^2 + 1)$. This means that $(x_0, y_0) \in A$ if and only if $(x_0, y_0) \in B$.

EQUALITY OF NUMBERS

1. Let's assume that $(x - y)^5 + (x - y)^3 = 0$.

Using the distributive property we can rewrite the equality as

$$(x - y)^3[(x - y)^2 + 1] = 0$$

The product of two or more real numbers is equal to zero if and only if at least one of the factors is equal to zero. Therefore either

$$(x - y)^3 = 0$$

or

$$[(x - y)^2 + 1] = 0$$

The first equality implies $x - y = 0$, or $x = y$. The proof is not complete because we have to show that this is the only possible conclusion.

The second equality can be rewritten as $(x-y)^2 = -1$. Since $(x-y)^2$ is always nonnegative, the equality will never be true.

Therefore the product $(x-y)^3[(x-y)^2 + 1]$ is equal to zero only if $(x-y)^3 = 0$, that is, when $x = y$.

We have to prove that if $x = y$, then $(x-y)^5 + (x-y)^3 = 0$. This is quite easy to do, indeed, in this case $x - y = 0$.

2. These two sequences are equal if $x^n = y^n$ for all $n > 1$.

Since $x^2 = y^2$, we obtain $(x-y)(x+y) = 0$. Therefore we have two possible conclusions: either $x = y$ or $x = -y$.

We can only accept $x = y$. Indeed, if we accepted the conclusion $x = -y$ we would obtain $x^3 = -y^3$. But $x^3 = y^3$ by hypothesis.

The converse of this statement is trivial.

3. By definition, a divides b if the division of b by a is exact (its remainder is zero). Therefore we can write $b/a = q$, with q a counting number.

Similarly $c/b = t$, with t a counting number, and $a/c = s$, with s a counting number. We can rewrite these equalities as:

$$b = aq \qquad c = bt \qquad a = cs \tag{\#}$$

If we use all of them we obtain

$$b = (cs)\,q = c(sq) = (bt)(sq) = b(tsq)$$

Therefore $b = b(tsq)$. Since b is a nonzero number, because 0 is not a counting number, we obtain $1 = tsq$. Since t, s, and q are all counting numbers, so in particular they are larger than or equal to 1, tsq can equal 1 only if $t = 1$, $s = 1$, and $q = 1$.

If we use this result in (#) we obtain

$$b = a \qquad c = b \qquad a = c$$

Therefore $a = b = c$.

4. Let a, b, and c be three counting numbers.

Set $d = \text{GCD}(ac,bc)$ and $e = \text{GCD}(a,b)$. We want to prove that $d = ce$.

1. $d \geq ce$

Since we can write $a = ke$ and $b = se$, we obtain

$$ac = k(ce)$$

$$bc = s(ce)$$

So ce is a common divisor of ac and bc. Thus $d \geq ce$ by definition of greatest common divisor.

2. $d \leq ce$

Since e is the greatest common divisor of a and b, we can write $a = ke$ and $b = se$ with k and s relatively prime.

So $ac = k(ce)$ and $bc = s(ce)$ with k and s relatively prime. This means that all the common factors of ac and bc are in ce.

Thus ce is larger than any other common divisor. So $d \leq ce$.

5. By hypothesis $(a/b)^m = n$, with n an integer.

Thus $a^m = b^m n$, or $a^m = b(b^{m-1}n)$. This means that b divides a^m.

Since a and b are relatively prime, their greatest common divisor is 1. Therefore b and a^m cannot have any factors in common other than 1.

Thus $b = 1$.

COMPOSITE STATEMENTS

1. We can rewrite the equation $x^2 = y^2$ as $x^2 - y^2 = 0$.

We can factor the difference of two squares and write the equation as

$$(x - y)(x + y) = 0$$

This equality implies that either $x - y = 0$ or $x + y = 0$.

If both x and y are equal to zero, both equalities are trivially true and $x = y$. Therefore let's assume that $x \neq 0$ and $y \neq 0$. From the previous equalities we obtain that either $x = y$ or $x = -y$.

If $x = -y$, since neither x nor y are equal to zero, we have that one of the two numbers is positive and the other is negative. But by the second part of the hypothesis the two numbers are nonnegative.

So it is not possible that $x = -y$. The only conclusion we can accept is $x = y$.

2. The function f is even. So $f(x) = f(-x)$ for all x in its domain.

But f is odd as well. So $f(x) = -f(-x)$ for all x in its domain. If we put together these two hypotheses we obtain

$$f(x) = f(-x) = -f(x)$$

Thus $2f(x) = 0$. This implies that $f(x) = 0$ for all x in the domain of the function.

3. Let n be a multiple of 3, and assume that n is not odd.
   Then n is even; that is, n is divisible by 2. Since n is divisible by 3 and 2, it is divisible by 6. So the two choices listed in the statement are the only possible ones. The statement is therefore true.

4. Let's assume that $x \neq y$. We can rewrite the equality $x^4 = y^4$ as

$$(x - y)(x + y)(x^2 + y^2) = 0$$

   Therefore either $x - y = 0$ or $x + y = 0$ or $x^2 + y^2 = 0$.
   The first equality implies $x = y$. But we have excluded this possibility. The second equality implies that $x = -y$. The statement is true if there are no more possible choices but this. The last equality is possible if and only if $x = y = 0$.
   Since we are working under the assumption $x \neq y$, we cannot accept this conclusion. So if $x \neq y$ the only possibility left is $x = -y$.

5. Let's assume that A is nonempty. The set $A - B$ is empty by hypothesis.
   By definition of the set $A - B$, this means that there is no element of A that does not belong to B. Then all the elements of A belong to B.
   Therefore $A \subset B$.

SOLUTIONS TO THE EXERCISES ON PAGE 76

Exercise 1

We are assuming that the two points, P and Q, are distinct. Therefore either the values of their x-coordinates or the values of their y-coordinates are different; that is, either $x_1 \neq x_2$ and/or $y_1 \neq y_2$.

Let's assume that $x_1 \neq x_2$ (geometrically this means that the points are not on the same vertical line). Since this implies that $x_1 - x_2 \neq 0$, we have

$$(x_1 - x_2)^2 > 0$$

The quantity $(y_1 - y_2)^2$ is always nonnegative (since it is the square of a number). Therefore

$$(x_1 - x_2)^2 + (y_1 - y_2)^2$$

is a positive number. This implies that d itself is a positive number.

This proves that the given statement is true under the assumption $x_1 \neq x_2$.

Similarly,we can prove the statement to be true if we assume that $y_1 \neq y_2$ (geometrically this means that the points are not on the same horizontal line).

The part of the proof for the y-coordinate is not a "must" because the formula used to evaluate the distance is completely symmetric with respect to the x- and y-coordinates (that means we could switch the two coordinates and the formula would not change); therefore whatever we say with respect to one of the coordinates is true for the other one as well.

There is a third part of the proof which again is not really needed, but we want to mention, just in case you thought about it.

It is possible to assume that $x_1 \neq x_2$ and $y_1 \neq y_2$ at the same time.

The given statement is still true under these hypotheses, because we have proved that it holds true when only one pair of coordinates has different values, which is a much weaker assumption than this last one.

Exercise 2

Let a be any real number and a' be an opposite of a. Then

$$a + a' = a' + a = 0 \tag{1}$$

Let b be another number such that

$$a + b = b + a = 0 \tag{2}$$

(Since we have two distinct numbers acting similarly on the number a, we should wonder how they interact. The answer to this question is not evident because we only know (1) and (2), and both equalities involve a. Then let's consider an algebraic expression that uses all the numbers we are using, namely a, b, and a'.

We must use addition, because it is the only operation used in the hypothesis. Let's try to start from

$$b + a + a'$$

Which conclusion can we reach?)

If we use the associative property of addition and (1), we obtain

$$b + a + a' = b + (a + a') = b + 0 = b$$

If we use the associative property and (2), we obtain

$$b + a + a' = (b + a) + a' = 0 + a' = a'$$

Therefore we have

$$a' = b + a + a' = b$$

or

$$a' = b$$

Thus the opposite of a is unique.

Exercise 3

a. By induction.

(i) The smallest positive integer number is 1.
Since $\ln 1 = 0$, it is true that $\ln 1 < 1$.
(ii) Let's assume that the inequality is true for n. Thus $\ln n < n$.
(iii) We have to show that the inequality holds for $n+1$; that is,

$$\ln(n+1) < (n+1)$$

(*Remember:* $\ln(n + 1)$ is not equal to $\ln n + \ln 1$.)
We have to try to use what we know about n and 1.

$$n + 1 = \frac{n+1}{n} n$$

Therefore

$$\ln(n+1) = \ln\left(\frac{n+1}{n} n\right) \qquad \text{properties of the function ln}$$

$$= \ln\left(\frac{n+1}{n}\right) + \ln n$$

$= \ln(1 + 1/n) + \ln n$ by inductive hypothesis $\ln n < n$

$< \ln(1 + 1/n) + n$ (#)

Since $n \geq 1$, $\frac{1}{n} \leq 1$. Therefore $1 + \frac{1}{n} \leq 1 + 1 = 2 \leq e = 2.72 \ldots$ So

$$1 + \frac{1}{n} \leq e$$

But ln is an increasing function; that is, the larger the number, the larger the value of its ln. Thus

$$\ln(1 + 1/n) \leq \ln e = 1$$

Then, if we use this in the chain of inequalities (#) we have

$$\ln(n + 1) < \ln(1 + 1/n) + n < 1 + n$$

or, equivalently,

$$\ln(n + 1) < n + 1$$

Now the proof is complete.

b. By graphing

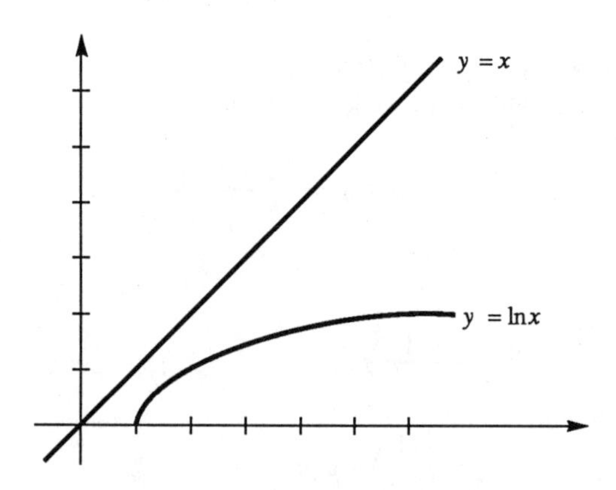

c. Let's consider the function $h(x) = \ln x/x$ for $x \geq 1$.

This function is never negative because $\ln x \geq 0$ for $x \geq 1$. Moreover, $h(1) = 0$.

We can either graph the function or use the first derivative test to check whether the function is increasing or decreasing and to look for maximum and minimum points (critical points).

$$h'(x) = \frac{\frac{x}{x} - \ln x}{x^2} = \frac{1 - \ln x}{x^2}$$

The function h has a critical point for the value of x such that

$$1 - \ln x = 0$$

or

$$\ln x = 1$$

Therefore

$$x = e$$

is the only critical point for h. Let's use the second derivative test to check which kind of critical point this is.

$$\begin{aligned} h''(x) &= \frac{\left(-\frac{1}{x}\right)x^2 - 2x(1 - \ln x)}{x^4} \\ &= \frac{-x - 2x(1 - \ln x)}{x^4} \\ &= \frac{-x(1 + 2 - 2\ln x)}{x^4} \\ &= \frac{2\ln x - 3}{x^3} \end{aligned}$$

Therefore $h''(e) = (2\ln e - 3)/e^3 = (2\text{-}3)/e^3 = -1/e^3$. This is a negative number. Thus e is a local maximum for the function.

So $h(x) \leq h(e)$ for all $x \geq 1$; that is, $h(x) \leq h(e) = \ln e/e = 1/e = 0.36\ldots$

Therefore $h(x) < 1$ for all $x \geq 1$. Since $\ln x/x < 1$, we have that $\ln x < x$ for all $x \geq 1$.

d. Let's consider the function $g(x) = \ln x - x$.

We know that $g(1) = -1$. Let's study the behavior of g using the first derivative test.

Since $g'(x) = -1 + 1/x$, it follows that g' is always negative for $x > 1$ and that $g'(1) = 0$.

Thus $g(1) = -1$ is the maximum value for g. This implies that $g(x) \leq -1 < 0$ for all $x \geq 1$. This proves that $\ln x < x$ for all $x \geq 1$.

Exercise 4

To prove that the sets are equal, we have to show that they have the same elements.

(i) $A \subset B$

Let x be an element of A. Then x is a multiple of 15; that is, $x = 15n$ for some integer number n.

Then x is a multiple of 5, because $x = (5)(3)\,n = 5(3n)$, and x is a multiple of 3 as well since $x = (3)(5)\,n = 3(5n)$. This means that x is an element of B.

(ii) $B \subset A$

Let y be an element of B, then y is a multiple of 3. So we can write $y = 3p$ for some integer number p. But y is a multiple of 5 as well. This implies that $3p$ is a multiple of 5. Since 3 is not a multiple of 5, p must be a multiple of 5. We can write $p = 5s$ for some integer number 5. Therefore $y = 3p = 3(5s) = 15s$.

Thus y is a multiple of 15 and it belongs to A.

The two parts, (i) and (ii), imply that A and B are equal.

b. (i) $A \subset B$

If *a* number is a multiple of 15, then it is *a* multiple of 3 and of 5, so it belongs to B.

(ii) $B \subset A$

This inclusion is not true. Consider the number 6. It is a multiple of 3, therefore it belongs to B. But it is not in A, because it is not a multiple of 15. Therefore A cannot contain the set B.

Moreover, because of part b (i), we know that A is contained in B. This means that A is a proper subset of B (that is, A is contained in B without being equal to B).

Exercise 5

(i) A solution exists.

$$ax = b$$

The number a is not equal to zero. Therefore it has a reciprocal: a^{-1}. Let's multiply both sides of the given equation by a^{-1}.We obtain

$$a^{-1}(ax) = a^{-1}b$$

By definition of the reciprocal of a number we obtain

$$1x = a^{-1}b$$

or

$$x = a^{-1}b$$

(ii) We can prove that the number $t = a^{-1}b$ is the unique solution of the given equation in two ways.

1. The number t is uniquely determined because of the uniqueness of the inverse of a.
2. Let s be another solution of the equation.

Thus $as = b$ by definition of solution. Therefore $at = as$. If we multiply both sides of the equality by a^{-1}, we obtain $s = t$.

This means that there is only one solution.

Exercise 6

1. If n^3 is an odd number, then n is an odd number.

(Since n^3 is an odd number, we can write it as $2q + 1$ for some q integer number. To find n we must take the cubic root of $2q + 1$. But there is no easy formula that allows us to calculate the cubic root of a sum. Therefore the fact that n^3 is an odd number does not help us reach the conclusion. Then we could try to prove the statement by contradiction.)

Let's assume that "n is not an odd number." The original statement is true if this assumption will contradict the hypothesis "n^3 is an odd number."

Since n is not an odd number, it must be even. Therefore we can write $n = 2p$ for some integer p. Then we have

$$n^3 = (2p)^3 = 8p^3$$

This equality implies that n^3 is an even number (it is divisible by 2). But this contradicts the hypothesis. Therefore n cannot be even, if n^3 is odd. Thus the original statement is true.

2. If n is an odd number, then n^3 is an odd number.
   If n is odd, $n = 2q + 1$ for some integer q. Therefore

$$n^3 = 8q^3 + 12q^2 + 6q + 1 = 2(4q^3 + 6q^2 + 3q) + 1$$

   The number $t = 4q^3 + 6q^2 + 3q$ is an integer.
   So we can write $n^3 = 2t + 1$. This means that n^3 is an odd number.

Exercise 7

(i) 1 implies 2

Suppose that the two inequalities in 1 are true. We can read them all together and obtain

$$a \leq b \leq a \qquad (\&)$$

Since the number a cannot be strictly smaller than itself, the chain of inequalities (&) can be true only if they are equalities. Therefore we have

$$a = b = a$$

or

$$a = b$$

Thus

$$a - b = 0$$

(ii) 2 implies 1

Since $a - b = 0$, we know that a and b are indeed equal. Therefore the inequalities

$$a \leq b \qquad b \leq a$$

are trivially true.

The proof is now complete.

Exercise 8

This is an existence and uniqueness theorem. Indeed, the statement can be read as:

1. Any nonzero number has a reciprocal. (This is an axiom.)
2. Such a reciprocal is unique.

Let's assume that there are at least two numbers, a^{-1} and s, with the properties

$$as = sa = 1 \quad aa^{-1} = a^{-1}a = 1$$

We want to prove that $a^{-1} = s$. Therefore we have to use the properties of these two numbers to try to compare them. Since

$$as = 1$$

and since $a^{-1} \neq 0$, we obtain

$$a^{-1}(as) = 1\,a^{-1}$$

Using the associative property of multiplication, we can rewrite the equality as

$$(a^{-1}a)\,s = 1\,a^{-1}$$

Therefore

$$1\,s = 1\,a^{-1}$$

or

$$s = a^{-1}$$

Therefore $s = a^{-1}$. So a has a unique reciprocal.

Exercise 9

Since the statement mentions "factors" and "division," we might consider writing p and q as products of prime factors. Thus

$$p = p_1 p_2 \ldots p_{r-1} p_r \quad \text{and} \quad q = q_1 q_2 \ldots q_{s-1} q_s$$

The numbers p_i need not be distinct. Similarly, the numbers q_i need not be distinct. But no p_i can equal any q_i. Using the factorizations above we have

$$p^n = (p_1)^n (p_2)^n \ldots (p_{r-1})^n (p_r)^n$$

Since the prime factors of p^n are equal to the prime factors of p, p^n and q have no common factors. Therefore q cannot divide p^n.

Exercise 10

Step 1: Is the formula true for $n = 1$, the smallest number we can use? When $n = 1$ we obtain

$$\frac{1}{1}\frac{1}{2} = \frac{1}{1+1}$$

Therefore the formula is true in this case.

Step 2: Assume that the formula is true for an arbitrary number n, that is,

$$\frac{1}{1}\frac{1}{2} + \frac{1}{2}\frac{1}{3} + \frac{1}{3}\frac{1}{4} + \ldots + \frac{1}{n}\frac{1}{n+1} = \frac{n}{n+1}$$

Step 3: Show that the formula is true when we use it for the next integer number, $n + 1$. So we have to prove that

$$\frac{1}{1}\frac{1}{2} + \frac{1}{2}\frac{1}{3} + \frac{1}{3}\frac{1}{4} + \ldots + \frac{1}{n}\frac{1}{n+1} + \frac{1}{n+1}\frac{1}{(n+2)} = \frac{n+1}{(n+1)+1}$$

or

$$\frac{1}{1}\frac{1}{2} + \frac{1}{2}\frac{1}{3} + \frac{1}{3}\frac{1}{4} + \ldots + \frac{1}{n}\frac{1}{n+1} + \frac{1}{n+1}\frac{1}{n+2} = \frac{n+1}{n+2}$$

Let's start the calculation.

$$\frac{1}{1}\frac{1}{2} + \frac{1}{2}\frac{1}{3} + \frac{1}{3}\frac{1}{4} + \ldots + \frac{1}{n}\frac{1}{n+1} + \frac{1}{n+1}\frac{1}{n+2}$$

associative property

$$= \left\{ \frac{1}{1}\frac{1}{2} + \frac{1}{2}\frac{1}{3} + \frac{1}{3}\frac{1}{4} + \ \dots \ + \frac{1}{n}\frac{1}{n+1} \right\} + \frac{1}{n+1}\frac{1}{n+2}$$

inductive hypothesis

$$= \frac{n}{n+1} + \frac{1}{(n+1)(n+2)} = \frac{n(n+2)+1}{(n+1)(n+2)}$$

$$= \frac{n^2+2n+1}{(n+1)(n+2)} = \frac{(n+1)^2}{(n+1)(n+2)}$$

$$= \frac{n+1}{n+2}$$

Since this is exactly the equality we were trying to prove, the formula is indeed true for all positive integer numbers.

Exercise 11

(The statement has only implicit hypotheses. Before proceeding we must be sure that we know the definition of rational number and the operations and properties of numbers.

We can then reformulate the statement as: If q is a rational number, then q cannot be equal to $\sqrt{2}$ (that is, $q^2 \neq 2$).

Since we cannot directly check the second power of all rational numbers, we will try to prove the statement by contradiction.)

Let's assume that there is a rational number $q = a/b$ with a and b relatively prime and $b \neq 0$ such that $q^2 = 2$.

Therefore $a^2 = 2b^2$. Since a^2 is a multiple of 2, a must be a multiple of 2. So $a = 2k$ for some integer number k. Then we have $4k^2 = 2b^2$. So $2k^2 = b^2$. Since b^2 is a multiple of 2, b must be a multiple of 2. So $b = 2s$ for some integer number s.

This means that a and b have 2 as a common factor. This is a contradiction because a and b had been chosen to be relatively prime.

Exercise 12

Let $y = ax + b$ and $y = cx + d$ be the equations of the two lines. Since the lines are distinct by hypothesis we know that either $a \neq c$ or $b \neq d$. The coordinates of the intersection points are the solutions of the system

$$y = ax + b$$

$$y = cx + d$$

Therefore we obtain $ax + b = cx + d$ or $(a - c)\,x = d - b$.

If $a = c$ (that is, $a - c = 0$), the system has no solutions because $b \neq d$. (We can explain this result geometrically. The two lines have the same slope. Therefore they are parallel and distinct. They have no points in common.)

If $a \neq c$ (that is, $a - c \neq 0$), we obtain $x = (d - b)\,/(a - c)$. Therefore the unique intersection point is the one having coordinates $x = (d - b)\,/(a - c)$ and $y = (ad - bc)\,/(a - c)$.

(This statement can be proved by contradiction as well. Start by assuming that the lines have two points in common, and use algebra to obtain a contradiction [that is, $a = c$ and $b = d$].)

Exercise 13

(Since it is impossible to check all negative numbers, we have to find a different way to verify whether the statement is true. Let's try the contradiction method.)

Assume that there exists a negative number z such that its reciprocal, z^{-1}, is not negative. By definition of reciprocal of a number $z(z^{-1}) = 1$. By the rules of algebra we know that z^{-1} cannot be zero.

Therefore z^{-1} must be positive. The product of a negative and a positive number is a negative number, and this generates a contradiction since 1 is a positive number.

Exercise 14

The conclusion has two parts:

1. The remainder is a number.
2. The remainder is equal to the number $P(a)$.

Since we are trying to evaluate the remainder of the division between the polynomials $P(x)$ and $x - a$, we should start from the division algorithm

$$\begin{array}{r|l} & q(x) \\ x - a & \overline{P(x)} \\ & \\ & \underline{r(x)} \end{array}$$

So we can write $P(x) = (x - a)\,q(x) + r(x)$.

The degree of the remainder must be smaller than the degree of the divisor, $x - a$, otherwise we could keep dividing. Since degree of $x - a$ is 1, $r(x)$ must have degree 0. So $r(x)$ is a number and we can write $r(x) = r$.

Therefore $P(x) = (x - a)\, q(x) + r$ where x represents any real number. We can now evaluate the expression for $x = a$ and obtain

$$P(a) = (a - a)\, q(a) + r = r$$

The proof is complete.

Exercise 15

There are several ways of proving that these three statements are equivalent. We will show that 1 implies 2, 2 implies 3, 3 implies 1.

1 implies 2: Since degree $P(x) \geq$ degree $(x - a) = 1$, the polynomial $P(x)$ can be divided by the polynomial $x - a$. Therefore

$$P(x) = (x - a)\, q(x) + r$$

But $P(a) = 0$. So $0 = P(a) = (a - a)\, q(a) + r = r$.

Since the remainder of the division is zero, $P(x) = (x - a)\, q(x)$. This means that $P(x)$ is exactly divided by $x - a$.

2 implies 3: By hypothesis the remainder of the division of $P(x)$ by $x - a$ is zero. So $P(x) = (x - a)\, q(x)$.

By definition this means that $x - a$ is a factor of $P(x)$.

3 implies 1: Since $x - a$ is a factor of $P(x)$, we can write

$$P(x) = (x - a)\, q(x)$$

Therefore

$$P(a) = (a - a)\, q(a) = 0$$

This proves that a is a root of the polynomial $P(x)$.

Exercise 16

By hypothesis the number

$$\lim_{x \to a} \frac{f(x) - f(a)}{x - a} f'(a) \qquad (*)$$

exists and it is finite. We want to show that $\lim_{x \to a} f(x) = f(a)$ or, equivalently, that $\lim_{x \to a} \{f(x) - f(a)\} = 0$.

Step 1: Is $\lim_{x \to a} \{f(x) - f(a)\} = 0$?

To reconstruct the fraction in (*), and therefore be able to use the hypothesis, we can divide and multiply by $(x - a)$. (It is algebraically correct to do so because $(x - a) \neq 0$.) In this way we obtain:

$$\lim_{x \to a} \{f(x) - f(a)\} = \lim_{x \to a} \frac{f(x) - f(a)}{x - a}(x - a)$$

$$= \lim_{x \to a} \frac{f(x) - f(a)}{x - a} \lim_{x \to a}(x - a)$$

$$= f'(a) \lim_{x \to a}(x - a) = f'(a)(0) = 0$$

Therefore $\lim_{x \to a} \{f(x) - f(a)\} = 0$.

Step 2: Does the result from step 1 imply that $\lim_{x \to a} f(x) = f(a)$?

By the properties of the limit,

$$\lim_{x \to a} \{f(x) - f(a)\} = \lim_{x \to a} f(x) - \lim_{x \to a} f(a)$$

The number $f(a)$ does not depend on x. Thus $\lim_{x \to a} f(a) = f(a)$. So

$$0 = \lim_{x \to a} \{f(x) - f(a)\} = \lim_{x \to a} f(x) - f(a)$$

Therefore $\lim_{x \to a} f(x) = f(a)$. So the function f is continuous at a.

Exercise 17

(This statement has only implicit hypotheses. Before proving it, we must be sure that we know when a number is said to be prime and when it is odd. Since there are infinitely many prime numbers, we cannot check that they are indeed all odd numbers. So we will try to prove the statement by contradiction.)

Let p be a prime number larger than 2. Let's assume that p is not odd. Then it must be even. So $p = 2n$ where n is some natural number. Therefore p can be divided by 2 and by n. This contradicts the fact that p is a prime number. Therefore p must be odd. (Be careful. Not all odd numbers are prime. Find an odd number that is not prime.)

Exercise 18

We can show that 1 is equivalent to 2 and that 2 is equivalent to 3.

1 implies 2
Let A^{-1} be the inverse of the matrix A. Let I_{2x2} be the 2x2 identity matrix. Then $A A^{-1} = I_{2x2}$. By the properties of the determinant,

$$\det(A A^{-1}) = (\det A)(\det A^{-1}) = \det I_{2x2} = 1$$

Therefore $\det A \neq 0$.

2 implies 1
We will explicitly find the matrix A^{-1}. Let

$$A = \begin{pmatrix} a & b \\ c & d \end{pmatrix} \qquad\qquad B = \begin{pmatrix} x & y \\ z & t \end{pmatrix}$$

We want to construct B such that $AB = I_{2x2}$.

From this equation we obtain a system with four equations and the four unknowns x, y, z, and t. Since the determinant of A is nonzero, we are able to solve the system and find

$$B = \frac{1}{\det A}\begin{pmatrix} d & -b \\ -c & a \end{pmatrix}$$

So B is the inverse matrix of A, since $AB = BA = I_{2x2}$.

2 implies 3
The system is formed by the two equations

$$ax + by = 0 \quad \text{and} \quad cx + dy = 0$$

If we solve it, we obtain $(ad - bc)\,x = 0$ and $(ad - bc)\,y = 0$. Obviously $x = y = 0$ is a solution.

Since $\det A = ad - bc \neq 0$, each one of these two equations has only one solution. So the system's only solution is $x = 0$, $y = 0$.

3 implies 2
The system is formed by the two equations

$$ax + by = 0 \quad \text{and} \quad cx + dy = 0$$

Since the system has a unique solution, either $a \neq 0$ or $c \neq 0$. (Otherwise the system would have an infinite number of solutions $(x,0)$, where x could be any real number.) Let's assume that $a \neq 0$.

Then $x = -by/a$. Substituting into the second equation we obtain

$$(ad - bc)\, y/a = 0 \quad \text{or} \quad (ad - bc)\, y = 0$$

In order for $y = 0$ to be the only solution of this equation we must have $ad - bc \neq 0$. So $\det A \neq 0$.

Exercise 19

This statement will be proved by induction.

a. Let's show that the formula is true when $k = 1$, the smallest number we are allowed to use.

$$1^3 = 1 \qquad 1^2(1 + 1)^2/4 = 4/4 = 1$$

So the formula works for $k = 1$.

b. By inductive hypothesis, we will assume that the formula works for the number $k = n$; that is,

$$1^3 + 2^3 + 3^3 + \ldots + n^3 = n^2(n + 1)^2/4$$

c. We want to prove that

$$1^3 + 2^3 + 3^3 + \ldots + (n + 1)^3 = (n + 1)^2[(n + 1) + 1]^2/4$$

We can use the associative property for the addition of numbers and rewrite the left-hand side of the equality to be proved as

$$[1^3 + 2^3 + 3^3 + \ldots + n^3] + (n + 1)^3$$

and use the inductive hypothesis from part b to obtain

$$[1^3 + 2^3 + 3^3 + \ldots + n^3] + (n + 1)^3 = n^2(n + 1)^2/4 + (n + 1)^3$$

$$= (n+1)^2[n^2/4 + (n+1)]$$

$$= (n+1)^2(n^2 + 4n + 4)/4$$

$$= (n+1)^2(n+2)^2/4$$

$$= (n+1)^2[(n+1) + 1]^2/4$$

Therefore the given formula is true.

Exercise 20

(The two numbers, a and b, appear in a formula. Therefore we can try to manipulate the formula to obtain explicit information about the two numbers.)

Since

$$ab = (a^2 + 2ab + b^2)/4$$

we have

$$4ab = a^2 + 2ab + b^2$$

or

$$0 = a^2 - 2ab + b^2$$

The right-hand side of the equality is equal to $(a-b)^2$. Therefore we obtain

$$(a-b)^2 = 0$$

This implies that

$$a - b = 0$$

So

$$a = b$$

Exercise 21

(We can start working on the given equation, which involves the two numbers a and b, hoping to get useful clues about them.)

From

$$ab = \frac{(a+b)^2}{2}$$

we obtain

$$2ab = a^2 + 2ab + b^2$$

Therefore

$$0 = a^2 + b^2$$

Since a^2 and b^2 are both nonnegative numbers, their sum can be equal to zero if and only if they are both equal to zero (since cancellation is not possible).

But $a^2 = 0$ and $b^2 = 0$ implies $a = b = 0$. This proves the statement.

Exercise 22

We are going to prove the given statement by induction.

a. The smallest number we can use is $k = 2$.

We have to add fractions whose denominators are integer numbers from 3 (which corresponds to $k + 1$) to 4 (which corresponds to $2k$). Therefore the left-hand side of the inequality becomes

$$\frac{1}{3} + \frac{1}{4} = \frac{7}{12}$$

Since $\frac{7}{12} > \frac{1}{2}$, the statement is true in this case.

b. Let's assume that the inequality is true for an arbitrary n.

c. Let's show that the inequality holds for $k = n + 1$,;that is,

$$\frac{1}{(n+1)+1} + \frac{1}{(n+1)+2} + \dots + \frac{1}{2(n+1)} > \frac{1}{2}$$

or

$$\frac{1}{n+2} + \frac{1}{n+3} + \dots + \frac{1}{2n} + \frac{1}{2n+1} + \frac{1}{2n+2} > \frac{1}{2}$$

The inductive hypothesis from part b can be rewritten as

$$\frac{1}{n+2}+\dots+\frac{1}{2n}>\frac{1}{2}-\frac{1}{n+1}$$

Therefore

$$\left(\frac{1}{n+2}+\frac{1}{n+3}+\dots\frac{1}{2n}\right)+\frac{1}{2n+1}+\frac{1}{2n+2}$$

$$>\left(\frac{1}{2}-\frac{1}{n+1}\right)+\frac{1}{2n+1}+\frac{1}{2n+2}$$

$$=\frac{1}{2}+\frac{1}{(2n+1)(2n+2)}>\frac{1}{2}$$

The statement is now proved.

Exercise 23

Since a is a multiple of b, we can write $a = bn$ for some integer n. Since b is a multiple of c, we can write $b = cm$ for some integer m.

We will now put all this information together to find a direct relation between a and c.

$$a = bn = (cm)\,n = cs$$

The number s is an integer number, because it is the product of two integer numbers, n and m. Therefore a is a multiple of c.

Exercise 24

The proof of this statement has two components:

1. If p is a nonzero rational number, then its reciprocal is a rational number.
2. If the reciprocal of the nonzero number p is rational, then the number itself is rational.

The reciprocal of a nonzero number p is the number q such that $pq = 1$.

Part 1: The number p is rational; therefore we can write $p = a/b$, with a and b nonzero relatively prime integer numbers. Therefore

$$pq = (a/b)\, q = 1$$

and

$$q = b/a$$

because $(a/b)(b/a) = 1$.

This means that q is rational.

Part 2: The number p^{-1}, which is the reciprocal of p, can be written as $p^{-1} = c/d$, with c and d nonzero integers. The number c cannot be zero, since p^{-1} is not zero. Then $p = d/c$ (because $(c/d)(d/c) = 1$), which proves that p is rational.

Exercise 25

Because a, b, and c are three consecutive integers, without loss of generality we can assume that a is the smallest of them and write $b = a + 1$ and $c = a + 2$. Then

$$a + b + c = a + (a + 1) + (a + 2) = 3a + 3 = 3(a + 1)$$

Therefore $a + b + c$ can be exactly divided by 3. The quotient of the division is $a + 1 (= b)$.

(Does this result relate in any way to the average of three consecutive integer numbers?)

Exercise 26

The proof is by induction.

a. We have to check whether the statement is true for $k = 1$.

$$k^3 - k = 1 - 1 = 0$$

and 0 is divisible by 3.

b. Let's assume that the statement is true for $k = n$; that is,

$$n^3 - n = 3p \qquad \text{for some integer } p$$

c. We want to prove that $(n + 1)^3 - (n + 1) = 3t$ for some integer t, using the information about $n^3 - n$.

$$\begin{aligned}(n+1)^3-(n+1) &= n^3+3n^2+3n+1-n-1\\ &= n^3+3n^2+3n-n\\ &= (n^3-n)+3(n^2+n)=3p+3q=3t\end{aligned}$$

Therefore the given statement is true.

The conclusions of Exercises 25 and 26 are similar. In each case we proved that a certain number ($a+b+c$ in Exercise 25 and k^3-k in Exercise 26) is divisible by 3. But the collections of numbers we can use are different. This is one of the reasons we choose different techniques.

We cannot use mathematical induction in exercise 25, because the three numbers could be negative. So there is not a smallest number for which to prove that the statement is true. If the statement were "Let a, b, and c be any three consecutive positive integers. Then 3 divides $a+b+c$," then we might have decided to use induction to prove it. Try this method and see what happens.

Exercise 27

We will prove this statement by contradiction. Let's assume there exists at least one sequence, $\{c_n\}$, such that $c_n > 0$ for all n and $L < 0$. By definition of limit of a sequence, for every $\varepsilon > 0$ there exists an N such that

$$|c_n - L| < \varepsilon \quad \text{for all } n > N\text{, which implies}$$
$$L-\varepsilon < c_n < L+\varepsilon \quad \text{for all } n > N$$

Consider $\varepsilon = -L/2$. Then there exists an M such that

$$L-(-L/2) < c_n < L+(-L/2) \quad \text{for all } n > M$$

Therefore

$$3L/2 < c_n < L/2 \quad \text{for all } n > M$$

This is impossible because $c_n > 0$ for all n, while $L/2 < 0$. Therefore $L \geq 0$.

We cannot say that the limit of a positive sequence must be positive. Indeed, let's consider the sequence $c_n = 1/n$. Every term of the sequence is positive, but the limit of the sequence is zero.

Exercise 28

Existence: Since $ad - bc \neq 0$, at least two of the numbers are not equal to zero. Let's assume that $a \neq 0$ and $d \neq 0$. Then from the first equation we obtain $x = (e - by)/a$. We can use this formula into the second equation and find that

$$y = (af - ce)/(ad - bc)$$

Therefore

$$x = (de - bf)/(ad - bc)$$

The two fractions are well defined since $ad - bc \neq 0$. So we have found one solution of the given system.

Uniqueness: This solution is unique because x and y are uniquely defined by the two equations written above.

(If you wish to do so, you can prove uniqueness in the following way: Let (x_1, y_1) and (x_2, y_2) be two solutions. Therefore

$$ax_1 + by_1 = e \qquad ax_2 + by_2 = e$$

and

$$cx_1 + dy_1 = f \qquad cx_2 + dy_2 = f$$

So

$$ax_1 + by_1 = ax_2 + by_2$$

$$cx_1 + dy_1 = cx_2 + dy_2$$

or

$$a(x_1 - x_2) = b(y_2 - y_1)$$

$$c(x_1 - x_2) = d(y_2 - y_1)$$

Therefore

$$x_1 - x_2 = b(y_2 - y_1)/a$$

and, from the second equation

$$(y_2 - y_1)(ad - bc) = 0$$

Since $ad - bc \neq 0$, the last equality implies $y_2 - y_1 = 0$; that is, $y_2 = y_1$. From this conclusion we obtain $x_1 = x_2$, as well. Therefore the two solutions are equal. This means that a unique solution exists.)

Exercise 29

This statement will be proved by induction.

a. The smallest number we can use is $k = 6$. In this case we obtain $2^6 = 64 > (6 + 1)^2 = 49$. So the inequality is true.

b. We will now assume that $2^n > (n + 1)^2$, and we will prove

c. $$2^{(n+1)} > \{(n + 1) + 1\}^2$$

$$2^{(n+1)} = 2^n 2 \qquad \text{by inductive hypothesis}$$

$$> (n + 1)^2 2 = 2n^2 + 4n + 2$$

We can evaluate $\{(n + 1) + 1\}^2$, and obtain

$$\{(n + 1) + 1\}^2 = n^2 + 4n + 4$$

Therefore we have to compare the quantities $2n^2 + 4n + 2$ and $n^2 + 4n + 4$. Since $n^2 \geq 36$

$$(2n^2 + 4n + 2) - (n^2 + 4n + 4) = n^2 - 2 > 0$$

If we put all the steps together we obtain

$$2^{(n+1)} = 2^n 2 > (n + 1)^2 2 = 2n^2 + 4n + 2$$

$$> n^2 + 4n + 4 = \{(n + 1) + 1\}^2$$

The inequality is therefore proved.

Exercise 30

This is an existence statement. So it is enough to find one number such that $2^k > (k + 1)^2$. Take $k = 6$.

Exercise 31

The statement does not seem to be true. We can try to show that it is false by counterexample. Take $t = 1$ and $q = 1/2$. Then $t + q = 3/2$, and this is not an irrational number.

(We can prove that the statement is false for every two rational numbers. Indeed, if t and q are rational, we can write $t = a/b$, with a and b relatively prime integers and $b \neq 0$, and $q = c/d$ with c and d relatively prime integers and $d \neq 0$. Then $bd \neq 0$

$$t + q = (ad + bc)/bd$$

So $t + q$ is a well-defined rational number and the given statement is false.)

Exercise 32

This is an existence statement. Therefore to prove it we can just show that there are three consecutive numbers whose sum is a multiple of 3; for example 3, 4, and 5.

The fact that this statement is true for any three consecutive integers (see Exercise 25) is irrelevant.

Exercise 33

The statement seems false. Therefore we should look for a counterexample. Take $n = 5$. Then 5 is a multiple of itself, but $5^2 = 25$ is not a multiple of 125.

Exercise 34

(We cannot use induction because we do not know what is the smallest number we can use.)

We are going to show that $n^2 + n = 2t$ for some integer t.
Using factorization, we obtain

$$n^2 + n = n(n + 1)$$

If n itself is an even number, then $n = 2q$ for some integer q. Thus $n^2 + n = 2q(n + 1) = 2t$.

If n is an odd number, $n + 1$ is an even number. So $n + 1 = 2m$ for some integer number m. This implies $n^2 + n = n(2m) = 2t$. Therefore the statement is true.

Exercise 35

Proof by induction.
a. Let's check the statement for $k = 6$.

$$6! = 720 > 216 = 6^3$$

So the statement is true in this case.

b. Let's assume that $n! > n^3$, and try to prove that

c. $$(n + 1)! > (n + 1)^3$$

$$(n + 1)! = (n + 1)\, n! > (n + 1)\, n^3 \geq (6 + 1)\, n^3$$

$$\geq n^3 + 6n^3 \geq n^3 + 6n^2 = n^3 + 3n^2 + 3n^2$$

$$\geq n^3 + 3n^2 + 3(6)\, n \geq n^3 + 3n^2 + 3n + 1 = (n + 1)^3$$

Therefore the statement is true.

Other Books on the Subject of Proofs

Copi, I. M. *Introduction to Logic*. New York: Macmillan, 1978.

Hermes, H. *Introduction to Mathematical Logic*. New York: Springer-Verlag, 1973.
Read Part I. Introduction.

Lucas, J. F. *Introduction to Abstract Mathematics*. Belmont, Calif.: Wadsworth, 1986.
Read Chapters 1 and 2.

Morash, R. P. *Bridge to Abstract Mathematics*. New York: Random House, 1987.
Read Chapters 1 to 5.

Polya, G. *How to Solve It*. Princeton, N.J.: University Press, 1945.

Polya, G. *Mathematical Discovery*. New York: Wiley & Sons, 1962.

Solow, D. *How to Read and Do Proofs*. New York: Wiley & Sons, 1982.

Smith, D.; Eggen, M.; and St. Andree, R. *A Transition to More Advanced Mathematics*, Belmont, Calif.: Wadsworth, 1983.

Stolyar, A. A. *Introduction to Mathematical Logic*. New York: Dover, 1970.
Read the first two sections.

Suppes, P. *Introduction to Logic*. New York: Van Nostrand, 1957.
Chapter 7 is quite interesting.

Whitehead, A. N. *An Introduction to Mathematics.* London: Oxford University Press, 1948.

Wickelgren, W. A. *How to Solve Problems.* San Francisco: Freeman, 1974.

An interesting and useful book for quickly checking mathematical terms is the well-known *Mathematics Dictionary* by Glenn James & Robert C. James, published by D. Van Nostrand Company, Inc. The booklet *Writing Mathematics Well* by Leonard Gillman (published by MAA) can be helpful too.

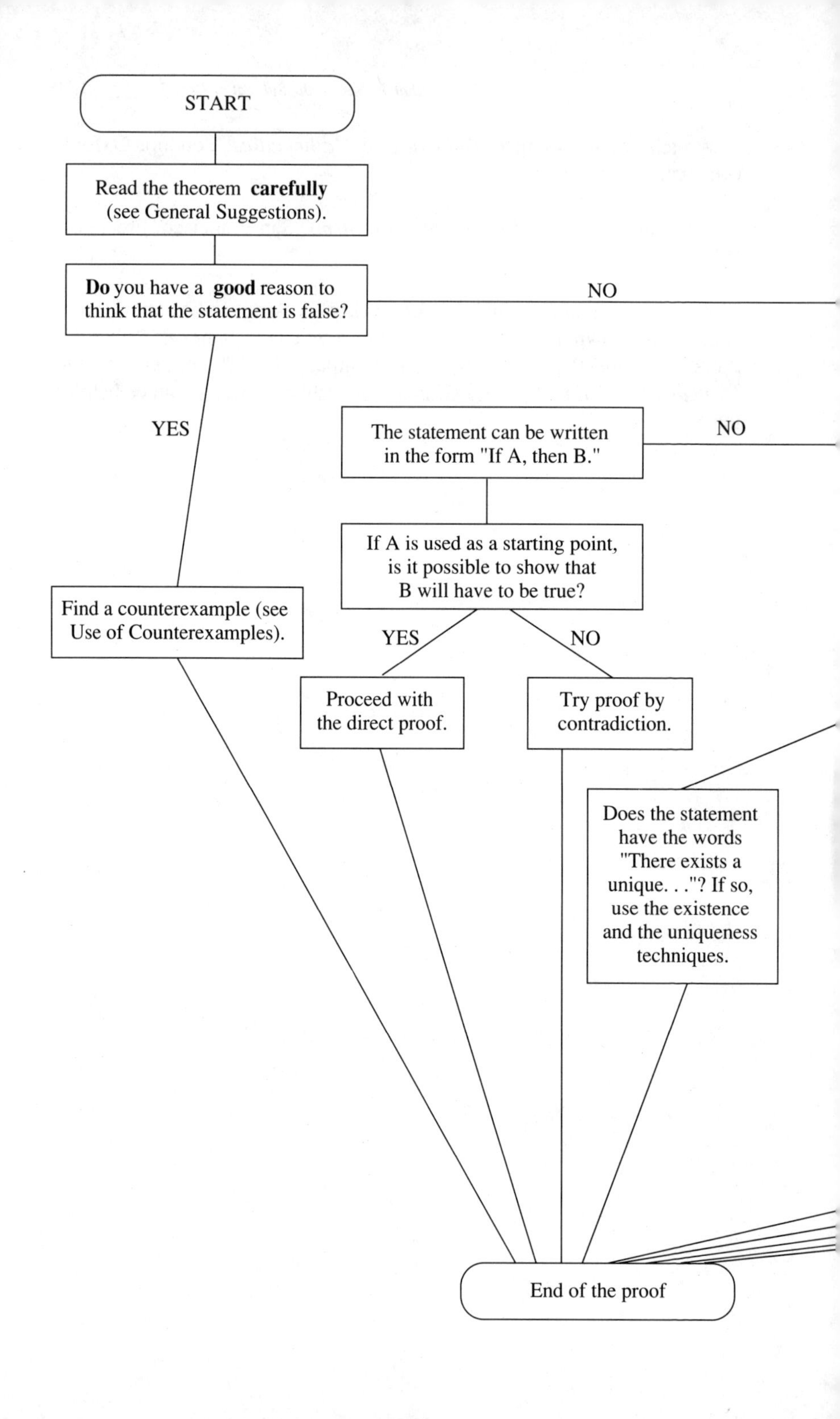
START
Read the theorem carefully (see General Suggestions).
Do you have a good reason to think that the statement is false?
NO
YES
Find a counterexample (see Use of Counterexamples).
The statement can be written in the form "If A, then B."
NO
If A is used as a starting point, is it possible to show that B will have to be true?
YES
NO
Proceed with the direct proof.
Try proof by contradiction.
Does the statement have the words "There exists a unique. . ."? If so, use the existence and the uniqueness techniques.
End of the proof

A Guide to Selecting a Method of Proof

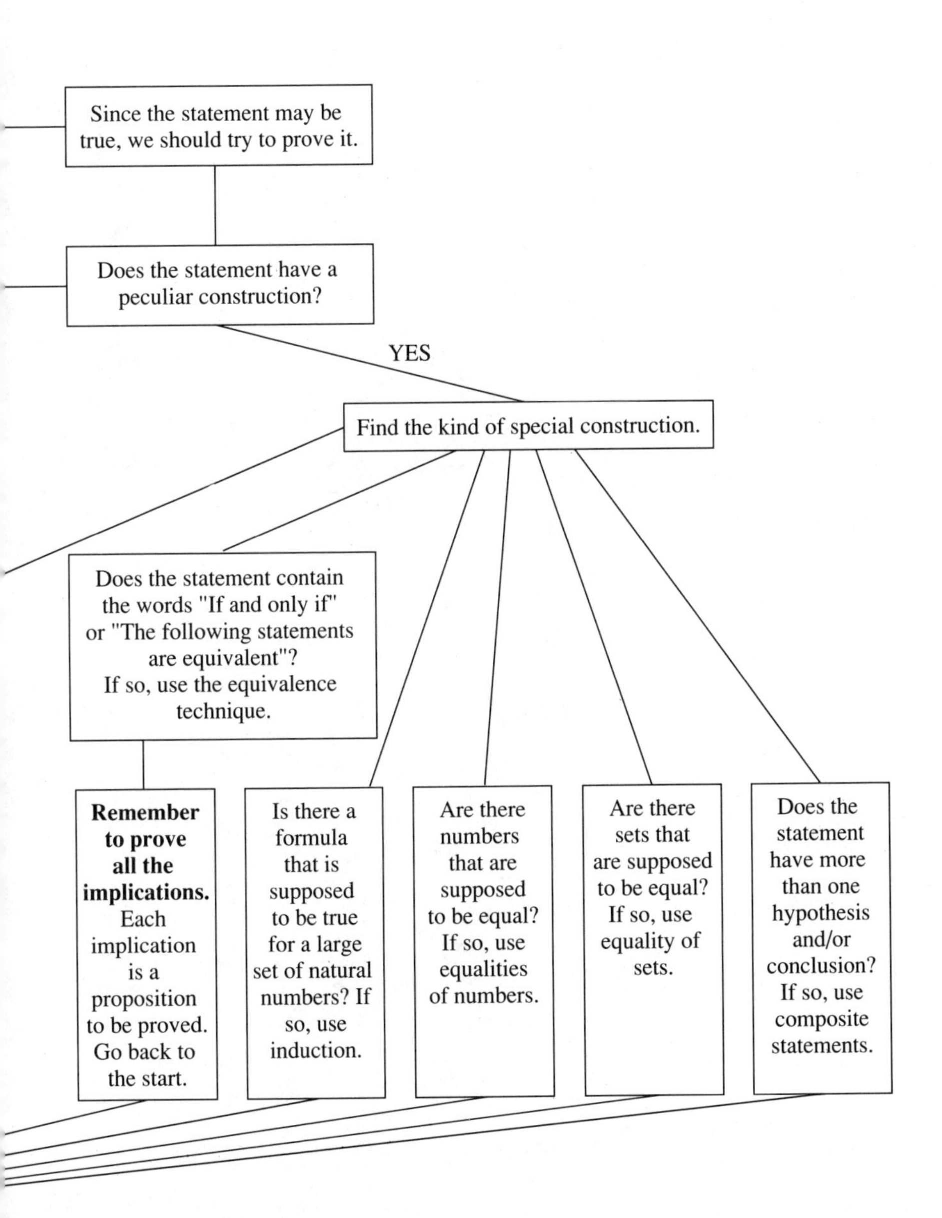